AF577240

Graphics and Animation in Surface Science

Graphics and Animation in Surface Science

Edited by

Dimitri D Vvedensky

The Blackett Laboratory and
Interdisciplinary Research Centre for Semiconductor Materials
Imperial College

and

Stephen Holloway

Surface Science Interdisciplinary Research Centre
University of Liverpool

Adam Hilger
Bristol, Philadelphia and New York

British Library Cataloguing in Publication Data

Vvedensky, Dimitri D.
Graphics and animation in surface science.
I. Title II. Holloway, Stephen
530.40285

ISBN 0-7503-0118-X

Library of Congress Cataloguing-in-Publication Data are available

Published under the Adam Hilger imprint by IOP Publishing Ltd
Techno House, Redcliffe Way, Bristol BS1 6NX, England
335 East 45th Street, New York, NY 10017-3483, USA
US Editorial Office: 1411 Walnut Street, Philadelphia, PA 19102

Printed in Great Britain by J W Arrowsmith Ltd, Bristol

Contents

Preface ix

Contributors xi

1 Introduction 1
- 1.1 Types of Visualization 2
- 1.2 Computational Modelling in Surface Science 3
  - 1.2.2 Ab-initio Total Energy Calculations 4
  - 1.2.2 Molecular Dynamics 5
  - 1.2.3 Monte Carlo Simulations 6
  - 1.2.4 Reactions at Surfaces 7
- 1.3 Outline of Book 8
  - References 9

2 The Use of High-Performance Graphics Workstations for Surface Science 11
- 2.1 Introduction 11
- 2.2 The Graphical User Interface 12
  - 2.2.1 Functionality of the Interface 12
  - 2.2.2 Usage of the Interface 13
- 2.3 What Can Computers Do? 15
  - 2.3.1 Computer Hardware 16
  - 2.3.2 Obtaining "Hardcopy" 18
  - 2.3.3 Rendering Techniques 19
- 2.4 Graphics in Surface Science 21
  - 2.4.1 The SYMM Package 22
  - 2.4.2 Properties—Using Existing Programs 23
- 2.5 For Example 23
  - 2.5.1 Magnesium Chloride—An Industrial Catalyst 23
  - 2.5.2 Transition-Metal Surfaces 24
  - Acknowledgement 24
  - References 24

3 Animation in Surface Science 26
3.1 Introduction 26
3.2 Brief Review of Computer Graphics 27
3.2.1 Graphics Libraries for Three Dimensions 30
3.2.2 Ray Tracing 30
3.3 Rendering Atoms for Solid-State Physics 31
3.4 The Animation Problem 33
3.4.1 Image Sequence 34
3.4.2 Image Compression 34
3.4.3 Storage and Display Box 37
3.4.4 Converting to Video Signals 38
3.4.5 Image Monitor 39
3.4.6 Video Animation Laboratory Block Diagram 39
3.5 Desktop Animation 40
3.6 Movietool: A Desktop Animation Tool 43
3.7 Conclusions 44
Acknowledgements 45
References 45

4 WINSOM Solid Modeller and its Application to Data Visualization 48
4.1 Introduction 48
4.2 Modelling Techniques 49
4.3 Input to WINSOM 50
4.4 Output from WINSOM 51
4.4.1 Static Pictures 51
4.4.2 Animation 52
4.4.3 Nongraphical Output 52
4.4.4 Low-Level Detailing 53
4.5 Applications 54
4.5.1 Molecular Modelling 54
4.5.2 Crystals 55
4.5.3 Liquid Crystals 55
4.5.4 Potential Fields 56
4.5.5 Biological Applications 57
4.5.6 Physics 57
4.6 Further Developments of WINSOM 58
4.6.1 WINSOM II 58
4.6.2 WINSOM90 59
4.6.3 Animation 59
Acknowledgements 60
References 60

5 Animations and Graphics in Molecular Dynamics Simulations 64
5.1 Introduction 64
5.1.1 The Purpose of Simulations 65
5.1.2 The Purpose of Graphics 65
5.1.3 Some Words of Caution 66
5.2 Programs 66
5.2.1 Setup 68
5.2.2 The Interaction Potential 68
5.2.3 The Principle of Molecular Dynamics 68
5.2.4 The Principle of the Monte Carlo Algorithm 68
5.2.5 Data File Format 69
5.3 Program Design 69
5.3.1 Development Tools 70
5.3.2 Language 70
5.3.3 Structure 70
5.3.4 Hardware 71
5.3.5 Examples of Analysis Programs 71
5.4 Animation 71
5.4.1 Correlation between Frames 72
5.4.2 Boundary Conditions 73
5.4.3 Economy of Plotting Commands 73
5.4.4 Size of Images 73
5.4.5 Playback Speed 76
5.4.6 Presentation on Paper 76
5.5 Design of Animation Process 76
5.5.1 Display of Atoms as Spheres 78
5.5.2 Display of Atomic Trajectories 83
5.6 Conclusions 83
Acknowledgements 84
References 84

6 Animation of Large-Scale Simulations 86
6.1 Introduction 86
6.2 Molecular-Beam Epitaxy 87
6.3 Modelling of MBE 89
6.3.1 Molecular Dynamics 89
6.3.2 Monte Carlo Simulations 90
6.3.3 Model for Si(001) Homoepitaxy 91
6.3.4 Epitaxial Growth on Vicinal Si(001) 93
6.4 Animation 95
6.4.1 The Problems We Face 96
6.4.2 The Problems We All Face 96
6.4.3 How to Animate 97

6.5 Animations of Si(001) Homoepitaxy 98
6.5.1 The Animated System 99
6.5.2 What Can Be Gained 99
6.5.3 Behaviour during Growth 100
6.5.4 Attainment of Equilibrium 101
6.5 Conclusions 102
Acknowledgements 102
References 102

7 Animation of Quantum Scattering Events using HyperCard 105
7.1 Introduction 105
7.2 Solution of the Schrödinger Equation 106
7.2.1 Dynamics through a Barrier 108
7.3 Steps to Generating a HyperCard Stack 113
7.3.1 A Walk through an Example 113
7.4 Closing Remarks 115
Acknowledgements 117
References 117

Preface

The nature and scope of computational studies of physical phenomena are changing rapidly, reflecting the advances in computer hardware and software that are making more powerful tools more generally available. The one-day conference on "Graphics in Surface Science" held in May, 1990 at the Surface Science Interdisciplinary Research Centre at the University of Liverpool had the aims of bringing together a group of people with an interest in using graphics and animation for exploring the physics of surface phenomena, to discuss and assess recent developments and the various implementations of graphics and animation that are currently available, and to highlight common problems and needs.

This volume was assembled with the goal of disseminating the information presented at the conference to a wider audience. The articles in this book are based upon talks given at the conference, but updated to include new developments and to provide introductory discussions of the graphical and computational methods used by each practitioner. Although the applications are presented for the most part in the context of surface phenomena, the graphical techniques that are discussed in these chapters have applications in wider contexts, and we hope readers in other fields will benefit from the following chapters.

We would like to express our gratitude to the staff of the Surface Science Interdisciplinary Research Centre at the University of Liverpool for their able assistance in the organization of this conference.

D.D. Vvedensky — London
S. Holloway — Liverpool

Contributors

Shaun Clarke (Chapter 6)
The Blackett Laboratory and Interdisciplinary Research Centre for Semiconductor Materials, Imperial College, London SW7 2BZ, United Kingdom

David M. Halstead (Chapter 7)
Surface Science Interdisciplinary Research Centre, University of Liverpool, Liverpool L69 3BX, United Kingdom

Nicholas M. Harrison (Chapter 2)
SERC Daresbury Laboratory, Daresbury, Warrington WA4 4AD, United Kingdom

Stephen Holloway (Chapter 1 and Chapter 7)
Surface Science Interdisciplinary Research Centre, University of Liverpool, Liverpool L69 3BX, United Kingdom

Ole H. Nielsen (Chapter 3)
Laboratory of Applied Physics, Technical University of Denmark, DK-2800 Lyngby, Denmark

M. William Ricketts (Chapter 4)
IBM UK Scientific Centre, Athelstan House, St. Clement Street, Winchester, Hampshire SO23 9DR, United Kingdom

Per Stoltze (Chapter 5)
Laboratory of Applied Physics, Technical University of Denmark, DK-2800 Lyngby, Denmark

Dimitri D. Vvedensky (Chapter 1)
The Blackett Laboratory and Interdisciplinary Research Centre for Semiconductor Materials, Imperial College, London SW7 2BZ, United Kingdom

Mark R. Wilby (Chapter 6)
The Blackett Laboratory and Interdisciplinary Research Centre for Semiconductor Materials, Imperial College, London SW7 2BZ, United Kingdom

1

Introduction

Dimitri D. Vvedensky and Stephen Holloway

Advances in scientific computation have led to a dramatic increase in the level of complexity and detail that can be treated by mathematical models. As architects of computer models have become more ambitious in their attempts to realistically represent physical systems, the need for ways to visualize the results of computations has become of paramount importance for representing and interpreting large amounts of data to develop a deeper and more intuitive understanding of computational experiments. It would not be an exaggeration to say that the presentation of data in a way that takes advantage of the human mind's capability to correlate images in space and in time is as important as the construction of the model itself.

There are becoming an increasing number of tools and techniques available for the presentation of scientific information. Aside from familiar examples such as patterns, intensity, and colour, computer-generated images are capable of incorporating sound, texture, and even feel. For example, an article on computer visualization contained in a recent issue of *Business Week* [1] reported a project at the University of Lowell where sound that can be traced to a particular location on the screen could be of use to air traffic controllers in monitoring crowded air space. In another project, at the Massachusetts Institute of Technology, a "joystick" is being developed that responds to chemical or magnetic forces. Thus, if molecules on a screen are put into an unfavourable position, the joystick offers resistance or "kicks back." These innovative uses of computer visualization underscore the impact that visualization can have on the presentation and interpretation of information.

Much of the current discussion concerning the application of computer graphics to scientific visualization originated in a report sponsored by the

National Science Foundation [2]. This report pointed out how computer graphics has the potential to revolutionize the interpretation of scientific data generated by experiments and simulations. Furthermore, the report recommended that funding organizations look into the possibility of funding research that would result in graphics software being made available to scientists.

The need for methodologies and tools for visualization across the disciplines of science, applied science, and technology, was also emphasized in a special issue of the journal *Computer* [3] devoted to visualization in scientific computing. Examples to illustrate the varied needs and uses of computer graphics were drawn from astrophysics, computational fluid dynamics, finite-element analysis of the stress distribution in a beam, molecular modelling, medical imaging, and mathematics. The theme for the issue was set by the quote "The purpose of computing is insight, not numbers" [4] in the introductory article by the guest editor.

1.1 TYPES OF DATA VISUALIZATION

Data visualization can take several forms, depending upon the application and the resources available. Two-dimensional x-y plots are among the simplest examples of visualization. These plots can aid in the interpretation of trends of functions of one or two variables, such as the reflected and diffracted intensity of an electron beam that is incident to a surface as a function of energy and beam direction. Such plots are useful for making quick visual comparisons between different sets of data, or between a theoretical prediction and an experimental measurement. Moreover, the graph of a function allows relations to be much more readily identified than simply perusing a set of complex mathematical formulae or columns of numbers.

More sophisticated renderings of surfaces with shadings and colour intensities are useful for displaying a function of two or more variables. The construction of surfaces with shading or with colour places greater demands upon graphics packages and upon the monitor, but is still a standard part of the scientific software in virtually all laboratories. Surface plots are used for many applications in surface science, ranging from the display of scans generated by scanning tunnelling microscopy to the potential energy surface seen by a diffusing atom on a surface.

An even more informative way to display the results of computational modelling is through animation. A sequence of images can provide a much more vivid illustration of how a system behaves than the individual presentation of the same static images. This is true not only where "time" is the variable that changes, such as in the simulation of time-dependent phenomena, but also in cases where the variable is a parameter whose effect upon a collection of data is sought. For such cases, animation provides a way to determine both the spatial and "temporal" correlations within a data set that is more direct than using correlation functions. In fact, viewing an animation can provide clues for identifying the important quantities that characterize the evolution of the system or the data. For systems that undergo complex dynamical behaviour or involve large numbers of atoms, animation also provides a simple way to follow individual trajectories, to observe the effect of fluctuations, the formation and decay of defects, and to identify the most active atoms.

However, animation also can place heavy demands upon computer resources both in terms of the processing power required to generate each frame, as well as the memory required to store the images. Furthermore, there is the additional complication of transferring the animation to a video tape for presentation to wider audiences, which requires specialized and expensive equipment. Nevertheless, graphics and animation are becoming increasingly popular for the presentation of scientific data, whether they are generated by simulations or by measurements. Some scientific conferences organize special sessions where videos are presented, and there are conferences devoted exclusively to the presentation of animations of computer-generated images [5].

The chapters in this book describe various applications of graphics and animation techniques to display theoretical and experimental data related to the study of surfaces. Surface science is concerned with the behaviour of surfaces, including structural, thermal and electronic properties, and their reactivity [6,7]. All of these properties are due ultimately to the interactions between electrons and nuclei near a surface and the response of these interactions to temperature, pressure, the introduction of atomic or molecular species, and various other external influences. A quantitative description of the physical properties surfaces beginning with this fundamental point-of-view is impractical, so most studies address individual aspects of surfaces. In the next section we will describe briefly the various types of computational modelling that are used to address specific features of surfaces.

1.2 COMPUTATIONAL MODELLING IN SURFACE SCIENCE

There are three techniques which are the most widely used for describing the behaviour of surfaces. These are the *ab-initio* total energy electronic structure calculations, molecular dynamics, and Monte Carlo simulations. Generally speaking, in moving from the *ab-initio* total energy calculations, to molecular dynamics and then to Monte Carlo simulations, the type of information that can be obtained becomes less specific, but the length and time scales over which the technique can provide information become greater. The *ab-initio* techniques provide detailed information regarding specific atomic configurations, and thus are best suited to characterizing barriers and pathways to diffusion [8] and other kinetic processes [9], and to determining the stability of relatively small collections of atoms.

The molecular dynamics method shares with *ab initio* total energy calculations the common feature that the Born-Oppenheimer approximation is used to separate the nuclear and electronic coordinates in the total Hamiltonian to obtain an effective Hamiltonian for the electronic coordinates. This Hamiltonian can then be used to obtain a potential energy surface for the nuclei as a function of their positions. In total energy calculations, this used to identify local minima in the total energy to obtain stable structures for a given configuration of atoms. In the molecular dynamics method, the expression for the total energy of the system as a function of the positions of the atoms is written as an expansion in terms of potentials, and the subsequent motion of the atoms is determined by the forces acting on the atoms. In the molecular dynamics method information concerning energy barriers for particular kinetic processes and the relative likelihoods of different events is a natural outcome of choosing a particular potential.

In the Monte Carlo method, the rate-determining events must be identified and rate constants must be estimated. The simplicity of the Monte Carlo method means that the details of local interatomic interactions are not explicitly incorporated into the model, but various processes are included on average through effective kinetic parameters. Thus, although this methods cannot be used to address effects that are too specific, comparisons with experiments are easier to make, because the simulations can be run under a greater variety of conditions. Furthermore, the Monte Carlo method provides a framework within which to identify the consequences of particular aspects of model potentials.

1.2.1 Ab-initio Total Energy Calculations

The most common approach for calculating electronic properties is to regard each electron as moving in an average potential generated by all the electrons. This potential, which must be calculated consistently, since the wave function of each electron affects the potential of all other electrons, is the sum of a Hartree potential and an "exchange-correlation" potential, which is a quantum mechanical correction to the Coulomb potential. The form of the exchange-correlation potential is not known exactly, so various approximations have been used [10]. Of these, the most common has been the local-density approximation, where the exchange-correlation potential of the interacting system is replaced by that corresponding to as homogeneous electron gas. Under favourable circumstances, the total energy can be calculated as a function of the position of the nuclei, and then minimized to obtain the ground state structure.

While applications to kinetic properties at surfaces are beginning to emerge for a few well-defined problems [8,11], the computational overheads for total-energy calculations have restricted applications mainly to static or quasi-static properties of surfaces. Even the computation of energy barriers to migration of a single atom on a surface represents a significant investment of computer resources, largely for three reasons. First, there is the reduced symmetry of a single atom on the surface. Second, a large number of trajectories must be sampled to determine the distribution of energy barriers. Third, for each trajectory. the local relaxation of the surrounding atoms must be taken into account, along with the redistribution of electronic charge resulting from the structural rearrangement. All of these factors conspire to make such calculations time-consuming and expensive.

Consequently, the computational modelling of time-dependent properties of surfaces has necessitated the sacrifice of some of the accuracy of the *ab-initio* methods for the sake of studying the dynamics of the system. The next two sub-sections describe the two main approaches used for studying the dynamics of surfaces.

1.2.2 Molecular Dynamics

In the molecular dynamics method, the expression for the total energy E of the system as a function of the positions of the atoms, $\mathbf{r}_1, \mathbf{r}_2, \ldots$ is written as an expansion in terms of n-body potentials:

$$E(\mathbf{r}_1, \mathbf{r}_2, \ldots) = \frac{1}{2} \sum_{i \neq j} V_2(r_{ij}) + \frac{1}{3!} \sum_{i \neq j \neq k} V_3(\mathbf{r}_{ij}, \mathbf{r}_{jk}, \mathbf{r}_{ik}) + \cdots$$

Here, $\mathbf{r}_{ij} = \mathbf{r}_i - \mathbf{r}_j$ are interatomic separations, and V_n represents an n-body interaction energy. We have assumed in writing (1) that we are dealing with a single-component system, so the interaction energies do not depend on the relative separations, not the atomic type. The two-body term represents the interaction of two atoms at positions $\mathbf{r}_i$ and $\mathbf{r}_j$ and depends only on the separation $\mathbf{r}_{ij}$. The three-body term V_3 depends upon the relative orientations of triplets of atoms, i.e., no simply upon interatomic distances but on bond angles as well. In order for the expansion of the potential in terms of n-body potentials to be meaningful, it must converge rapidly with increasing n; in fact, virtually all studies truncate this expansion at either the second or the third order. Given the potential, the Hamiltonian of the system can be constructed from which the time-development of the system can be calculated from Hamilton's equations

The two important features of the molecular dynamics method are the specification of potential and the time step used for the integration of the equations of motion. The potential is determined by a combination of semi-empirical fitting to observed behaviour combined with physically-motivated characteristics. For example, a two-body potential is appropriate for systems where the formation of chemical bonds is not an important feature of the dynamics, such as the interaction of He with a metal. However, for most systems, a three-body component is also required to describe the interatomic interactions responsible for bond-formation. Once the potential has been chosen, and the interaction energy determined, the forces acting on each atom is calculated and, at each time step, the atoms are moved in the direction of the forces. This method can be used both to calculate the structure of a collection of atoms by determining the minimum energy as a function of the atomic positions and to determine the dynamics of a system, such as the melting or freezing of a surface.

The commitment to a fixed interatomic potential has both advantages and disadvantages. On the positive side, the description of dynamic phenomena is enormously simplified in comparison with *ab initio* methods, since the individual potentials need not be recalculated as the atomic configuration is varied. For systems where hybridization is not a strong function of the local environment, this is a good approximation. On the other hand, the calculated behaviour of systems which do show strong effects of rehybridization, such as C and Si, different potentials will produce different relative energies among different structures, which then can lead to different descriptions of the dynamics.

The second important feature of the molecular dynamics method is the time step used between successive evaluations of the forces acting on the atoms and the corresponding adjustments of the atomic positions. Since the all of the dynamical detail of the system is included in this method, this time step must describe the vibrational properties of the individual atoms. Since the atomic vibrational period is of the order of 10^{-13} seconds, this gives an upper bound to the time step. The magnitude of the time step thus determines the "real time" for which the simulation is allowed to develop. In particular, phenomena that occur over a time scale of seconds will present problems in terms of computational resources for conventional molecular dynamics.

1.2.3 Monte Carlo Simulations

The Monte Carlo method is an additional level of abstraction over the molecular dynamics method. The effect of fast dynamical events is taken in account phenomenologically through transition rates for slower events. For example, in describing the mobility of surface adatoms, the diffusion can often be approximated as a nearest-neighbour hopping process with a transition rate given by the product of an attempt rate, which is typically of the order of the atomic vibrational frequency, and the probability of success per attempt, which is represented as an exponential involving the energy barrier to the process. The hopping of atoms is then described by comparing the probability of the hopping, p, with random number, n, chosen from the interval $[0, 1]$: the hopping occurs only if $n \leq p$. Other kinetic events are similarly treated.

While the details of the underlying mechanism for the hopping are lost, the effect of the fast processes is correct on average. Thus, if the large-scale features of the dynamics of a system are of interest, rather than the details of the structure, the Monte Carlo method can offer considerable advantages over the molecular dynamics method, both in terms of the "real time" over which the simulation evolves, as well as the number of atoms included in the simulation.

It must be emphasized that the construction of a model for a Monte Carlo simulations can often be greatly simplified and justified by appealing to a related molecular dynamics simulation. This includes the identification of the important physical process, as well as the numerical values of the corresponding kinetic barriers.

1.2.4 Reactions at Surfaces

Much of the early work aimed at understanding the behaviour of surfaces primarily involved the structural, electronic, and vibrational properties of clean and absorbate-covered surfaces. That work has matured to the point where the structure of quite complex systems can be analyzed and where the dynamical behaviour of some systems can be modelled. A natural next step is the study of dynamic and kinetic phenomena involving interactions between atoms molecules and surfaces. The interaction of molecules with surfaces are of evident technological importance, including processes such as catalysis, corrosion, and chemical vapour deposition. This work builds upon the expertise gained in the theoretical and experimental study of gas-phase reactions, and includes methods such as molecular-beam scattering and laser pumping and diagnostic techniques to study the interaction of atoms and small molecules with well-characterized metal surfaces. Quantities that can now be measured include the vibrational and translational energy dependence of dissociative chemisorption at metal surfaces and the internal energy distributions of scattered and desorbed molecules.

The modelling of the interactions between molecules and surfaces involves considering many physical processes, including the dynamics of the collision between the molecule and the surface, the dissipations of the energy of the incident molecule at the surface, the chemisorption and possible dissociation of the molecule, and the migration of the fragments along the surface. The complexity of the possible pathways for molecule-surface interactions has made it difficult to apply conventional theoretical techniques that are used at surfaces, because in many ways, methods are required that incorporate features both from atomic and molecular studies on one hand, and solid-state studies on the other. Nevertheless, a consistent conceptual framework is beginning to emerge, at least for relatively simple systems, based upon quantum mechanical calculations of varying degrees of complexity, ranging from *ab initio* total-energy calculations, to the integration of the time-dependent Schrödinger equations for the evolution of a wavepacket on a potential-energy surface (for a review, see Ref. [12]).

1.3 OUTLINE OF BOOK

As the preceding section shows, the complexity and diversity of surface phenomena is reflected in the computational techniques used to model these phenomena. The papers in this volume reflect this diversity and address the graphical requirements and the different types of implementations of visualization that have been used.

The first three chapters are devoted to the techniques of visualization in three different settings. Chapter 2 by Nicholas Harrison discusses the implementation of graphics workstations to assemble and present different types data, both experimental and theoretical, working in a major central facility (Daresbury Laboratory). This may involve representing the idealized structure of a surface in terms a "ball-and–stick" model, various types of spectra, such as x-ray-absorption, low-energy electron-diffraction, as well the results of total-energy calculations, including the total charge density and the optimised structure. The article focuses upon the *Graphical User Interface*, as the basis of an integrated laboratory-wide network that gives a standardized access to machines, data, and computational tools.

Chapter 3, by Ole Nielsen, provides an introduction to the terminology and concepts of graphics and animation and describes applications to surface science within the environment of a scientific laboratory. Different ways of rendering the atomic structure of surfaces are discussed, including the advantages and disadvantages of each type in terms of utility and computational requirements. The processes involved in animating sequences of images are described, from the rendering stage, through the storage of the images, to the conversion to video signals. The chapter concludes with a description of *Movietool*, an animation control unit developed by the author.

Chapter 4, by Bill Ricketts, describes a general-purpose solid modelling package called WINSOM developed at the IBM UK Scientific Centre at Winchester. This chapter provides a discussion of the mechanics involved in the design and assembly of a graphics package intended for general distribution. Also included is a brief description of some of the many applications of WINSOM in physics, biology, and chemistry. An interesting application of WINSOM, quite apart from scientific visualization, is its use as an artistic medium. Striking and realistic images have been generated by using the features of WINSOM as a palette. The incorporation of some of the other senses into computer graphics, as discussed in the Introduction, holds promise of some intriguing artistic possibilities.

The chapter by Per Stolze (Chapter 5) is the companion to the chapter by Nielsen and reviews the application of the methods of Chapter 2 to animating molecular dynamics simulations. This chapter discusses several method used to follow the motion of individual atoms in molecular dynamics simulations, including labelling snapshots of labelled atoms, and trajectories of all atoms. Each representation of the time development of a simulation has its uses. For example, trajectories show how likely atoms are to move from their initial positions, and emphasise the regions of greatest activity. Labelling individual atoms, on the other hand helps to identify pathways for diffusion. Animations have the versatility to show both aspects of the problem, with the animation rate providing the control to emphasize one aspect or the other.

The chapter by Mark Wilby and Shaun Clarke (Chapter 6) describes the application of the WINSOM solid modeller to animations of large-scale simulations of lattice models of growth. Some of the problems presented by these animations are the same as those discussed by Stolze in Chapter 5, but the animation of simulations of tens of thousands of atoms presents more specific problems simply because of the enormity of the data involved. This Chapter first introduces the technique of molecular-beam epitaxy, which is one of the most common methods used to study the growth of metals and semiconductors, and then describes the model for the growth kinetics. The emphasis of this Chapter is on the growth on Si(001), and the ways in which the animation of the simulation has clarified some of the intriguing features of the ways this system grows.

Where Chapters 5 and 6 discuss the application of sophisticated software to producing very detailed animations of dynamical behaviour near surfaces, the chapter by Halstead and Holloway (Chapter 7) describes the animation of wave packet scattering using the one-dimensional time-dependent Schrödinger equation using a desk-top computer. The model was originally developed to study certain aspects of the interaction of simple molecules with surfaces, but can also be used to illustrate general features of the scattering of a wave packet by a potential barrier.

REFERENCES

1. *Business Week*, Nov. 12, 1990, p. 77.
2. B.H. McCormick, T.A. DeFanti, and M.D. Brown, eds., "Visualization in scientific computing," *Computer Graphics* **21**(6), Nov. 1987.
3. *Computer*, **22**(8), Aug. 1989.
4. R.W. Hamming, **Numerical Methods for Scientists and Engineers** (New York, McGraw-Hill, 1962).

5. A recent example is the *Second Eurographics Workshop on Animation and Simulation* held in Vienna on 1-2 September 1991.
6. A. Zangwill, **Physics at Surfaces** (Cambridge, Cambridge University Press, 1988).
7. V. Bortolani, N.H. March, and M.P. Tosi, **Interactions of Atoms and Molecules with Solid Surfaces** (New York, Plenum, 1990).
8. P.J. Feibelman, "Diffusion path for an Al adatom on Al(001)," *Phys. Rev. Lett.* **65**, 729 (1990).
9. E. Kaxiras, O.L. Alerhand, J.D. Joannopoulos, and G.W. Turner, "Microscopic model of heteroepitaxy of GaAs on Si(001)," *Phys. Rev. Lett.* **62**, 2484 (1989).
10. S. Lundqvist and N.H. March, eds. **Theory of the Inhomogeneous Electron Gas** (Plenum, New York, 1983).
11. P.J. Feibelman, "Orientation dependence of the hydrogen molecule's interaction with Rh(001), *Phys. Rev. Lett.* **67**, 461 (1991).
12. S. Holloway, "Theory of adsorption-desorption kinetics and dynamics," in Ref. [6], pp.567-598.

2

The Use of High-Performance Graphics Workstations for Surface Science

Nicholas M Harrison

The last two years have seen considerable efforts at Daresbury Laboratory to develop a research environment which makes optimal use of both existing software and new graphics hardware. The existing software consists of many widely-used codes (usually written in FORTRAN) which have been developed over the last twenty years by various university groups and Daresbury staff. The packages relevant to surface science have many functions but are basically concerned with either the collection and analysis of data from experiments on the synchrotron or direct theoretical treatment of the physical and chemical properties of surfaces. This article describes the application of high-performance graphics workstations to surface science based upon research environment at Daresbury.

2.1 INTRODUCTION

Over the last decade the explosion in the application of computers to the problems of data acquisition and processing, as well as the theoretical calculation of surface properties, has lead to ever increasing demands for computer power. In addition to this, the wide variety of theoretical and experimental tools in use require more flexible access to various computer functions, such as database systems and visualisation tools, as well as basic number-crunching. This explosion in demand for resources is matched by the developments in computer hardware and software. The fast growth in low-cost processing and graphical performance, coupled with some standardisation of the underlying software, means that one is now able to link together machines of various functionality to produce an efficient and flexible working environment. This article is concerned primarily with the application of graphics workstations to surface science but is based on our experience of developing such an environment at Daresbury Laboratory.

A large amount of application software specific to surface science problems has been developed over the past 20 years, much of which has been compiled and is supported by the *Collaborative Computational Project* in surface science (CCP3) [1]. Applications which allow one to describe and edit surface structures and to calculate spectroscopic and energetic properties of surfaces at many different levels of approximation are available. In surface science it is often the case that a specific problem may not be solved without resorting to a variety of theoretical techniques and experimental probes. This fact, one of the motivations for forming CCP3, means that the development of a common user interface to surface science application codes has been a central feature of our use of graphics workstations.

2.2 THE GRAPHICAL USER INTERFACE

A partial realisation of this *Graphical User Interface* has been achieved at Daresbury Laboratory over the past 2 years on a network of UNIX machines including compute servers, (such as a Convex C220, a CRAY X-MP 4/8 [2], and an Intel IPSC/860 parallel supercomputer) graphics workstations and database servers. Underlying software has included a graphical interface and rendering program under development at Daresbury, DISPLAY [3], and the commercially available systems *A*nimation *P*roduction *E*nvironment (apE) [4] and *A*pplication *V*isualization *S*ystem AVS [5]. A new software system similar in concept to apE and AVS, called Explorer, has been announced by Silicon Graphics.

2.2.1 Functionality of the Interface

The basic functionality provided by these systems is listed below:

Spawning of applications. One must be able to initiate an application on an appropriate computer. This function is widely available on UNIX machines via the “remote shell” facility.

Passing data between applications. The ability to move data between applications, which may be running on separate machines, is vital in the heterogeneous computing environment brought about by the development of efficient network protocols and multi-processor computers. The flow of data may be achieved in several different ways. A widely-used and easy-to-implement solution is to write and read disk files. For inter-machine communication one then has the problem of moving files between machines, which is facilitated by protocols such as *F*ile *T*ransfer *P*rotocol FTP [6] or *N*etwork *F*ile *S*ystem NFS [7]. For applications requiring sophisticated dialogue or efficient communications, file-based transfer is very cumbersome. In a UNIX environment applications can open two-way communications channels called *sockets* or *streams* which may be used to pass data within a machine or via a network subject to the user handling any incompatibilities between the binary formats used on different computers. This is a flexible and efficient way of communicating between machines, but within one machine a better method is to pass references to areas of *shared memory*. Fortunately, the ubiquitous nature of this problem has led to the production of both academic [8] and commercial [9] software which provides for direct inter-process communication in a hardware-transparent fashion.

Rendering facilities. The translation of data into graphical objects and their subsequent display on a variety of devices (screen, plotters, video, *etc*). From simple line graphs to coloured isovalue surfaces the algorithms for performing such tasks are widely available.

User interaction. The communication between a user and application codes is usually driven via a series of menus or mouse-selected buttons and sliders. Toolkits for providing such "interactives" are also widely available.

With these basic features a more of less sophisticated user interface may be constructed. A simple conception of the rôle played by such an interface is displayed in Figure 2.1.

2.2.2 Usage of the Interface

The above discussion has been a little abstract; as a concrete example let us consider a typical scenario. We are to perform a series of experiments designed to understand the structure of a given surface. There are many measurements which one might perform. Low-angle X-ray diffraction will provide a surface sensitive diffraction pattern, as would low-energy electron-diffraction (LEED). Local structural information would be obtained from X-ray absorption experiments (XANES) and more chemistry sensitive information from ultra-violet photoelectron spectroscopy (UPS) and angle-resolved ultra-violet photoelectron spectroscopy (ARUPS). A certain amount of information may be extracted directly from the spectra—all we need to do is collect, label and plot them. A great deal more can be learned by analysing the spectra in terms of the various theoretical models available to us. Firstly, one might directly calculate the theoretical spectra which correspond to various likely atomic arrangements; these may be directly compared to measurements and will act as a guide in choosing other likely atomic arrangements. One might also try to predict the surface structure from a detailed treatment of the surface chemistry. As

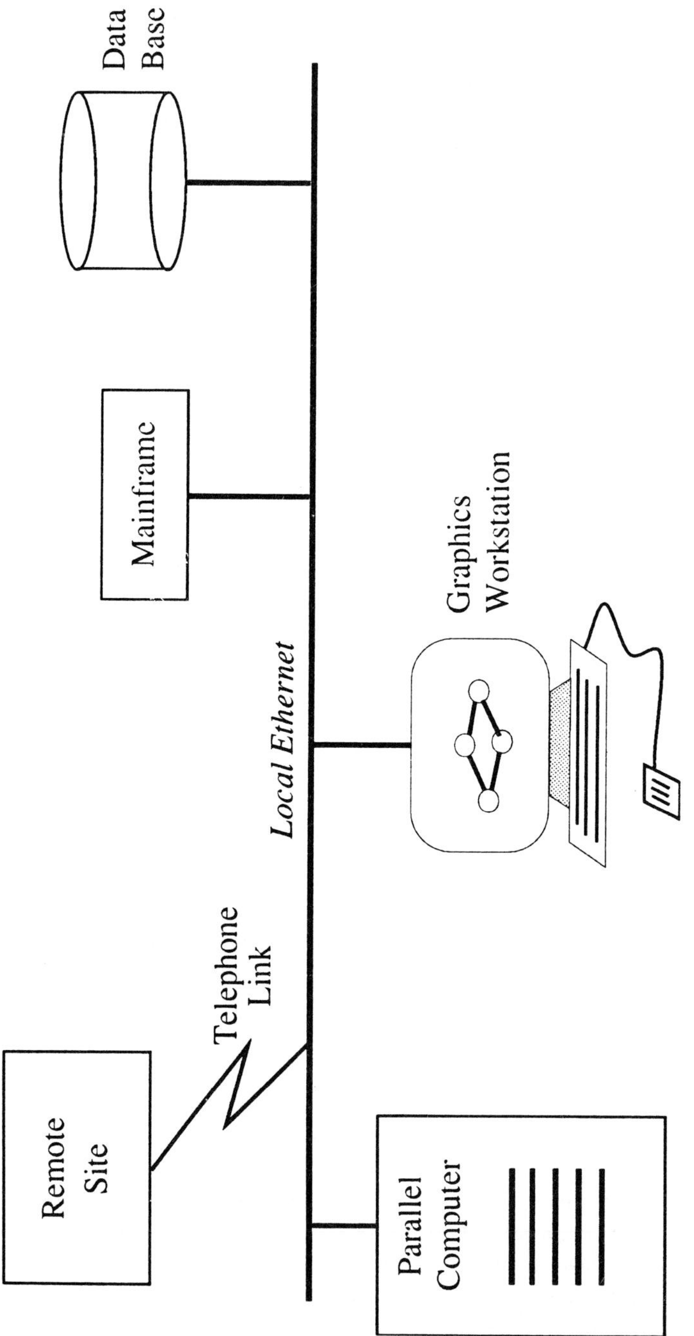

Figure 2.1 The relationship between the workstation, the local network and remote sites.

mentioned above, packages already exist to perform each of the basic calculations needed.

It is clear that to perform this analysis is an extremely time-consuming exercise which requires a large degree of technical competence and familiarity with the details of many different computational packages. Another complicating feature is the different time scales on which calculations may be performed. From calculating a simple qualitative X-ray diffraction pattern to a full-blown quantum-mechanical treatment of the surface chemistry takes us through many orders of magnitude of computing power; from an IBM-PC to a CRAY X-MP.

In this instance the role of a graphical user interface is clear. The *physical* problem is specified by the atomic structure of the surface. A graphical editor, using simple "ball-and-stick" representations, can be used to build a prototype structure. This information is now available, in an appropriate format, to any application code on any hardware attached to the network; the problem of formatting the information in a manner suitable for a specific application (*filtering*) is attacked as part of the user interface. An application which requires extra information (control of the computational algorithms, for example) obtains such information via the user interactives provided within a communications library to which it, or its filters, may be linked. Any data produced by the application is simply posted back to the user interface and enters the interactive domain of the user. It may then be passed to appropriate renderers or forwarded to another application code. Experimental data can be accessed directly from data-acquisition systems attached to experimental equipment or from database facilities.

Thus, a wide variety of existing packages are brought within the control of a single user interface; the investment of time in converting packages to use standard formats and to conduct dialogues in order to determine operating conditions is easily justified in terms of the time saved on technical aspects of driving an application code.

In the remainder of this article I will review the available computer hardware for performing high-performance three-dimensional graphics, and outline the concepts needed to use it effectively. The requirements of surface science are then examined with particular examples where the methodology described above has been put into practice.

2.3 WHAT CAN COMPUTERS DO?

The revolution in computer graphics which has occurred over the past ten years has been driven by developments of the basic computer hardware and the standardisation of much of the essential software tools. In this section I will explain the essential features of a modern high performance workstation and discuss the ways in which one may take advantage of them.

2.3.1 Computer Hardware

On almost all modern computers the screen is a *raster* display [10] rather similar to a colour TV set. The screen consists of picture elements (pixels) each of which is able to glow red, green and blue independently. An electron beam from a cathode ray tube is used to scan every pixel on the screen and to light-up the required red, green, and blue components to form the image. Typically, a screen will contain about 1000×900 pixels and will be completely redrawn 60 times per second. The information about which pixels are to be lit and what colours they should have is contained in fast memory called a *frame buffer*. The rendering of a two-dimensional object is simply a matter of decomposing the image into its individual colour components and then lighting the appropriate pixels.

Considerably more difficulties arise when we wish to represent a three-dimensional object on the two-dimensional screen. The faithful representation of a three-dimensional object requires a fairly sophisticated treatment of the relationship between the surface properties of the object, the way it is lit and the direction from which it is viewed. This task may be performed with varying degrees of accuracy but the computer must perform a sequence of basic steps. First, a *clipping* process is applied. From the direction of view and the *field of view* the parts of the object which may not be "seen" are discarded. The optical properties of the visible surface (colour, reflectivity, specularity, *etc.*) are now used with the nature of the illumination (spot, diffuse, coloured, *etc.*) and the viewing angle to perform a *lighting calculation*. This may be as simple as determining the local colour of polygons used to represent a surface or may involve more sophisticated techniques such as *smooth shading* the polygons or tracing the reflections undergone by individual rays of light (*ray tracing*). The final result of this calculation is a two-dimensional list of coloured pixels in the frame buffer which may be loaded into the screen in 1/60th of a second.

The implementation of algorithms for such calculations is a delicate balance between a quality and expense (in computer time and memory). One of the major advances of the past few years is that rather sophisticated algorithms have been implemented by computer manufacturers and supplied as callable libraries. Also many machines are now available which contain special hardware designed to perform the above operations (or machine-coded software on fast, dedicated processors). On a modern workstation one expects an image consisting of surfaces specified in terms of ten thousand square patches to be clipped, lit and shaded and redrawn at a rate of 10–20 times per second. This allows one to generate complex images and to rotate them interactively on the screen. The relationship of application code to the interface, hardware and end product are summarised in Figure 2.2.

As well as providing high performance, the use of specialized graph-

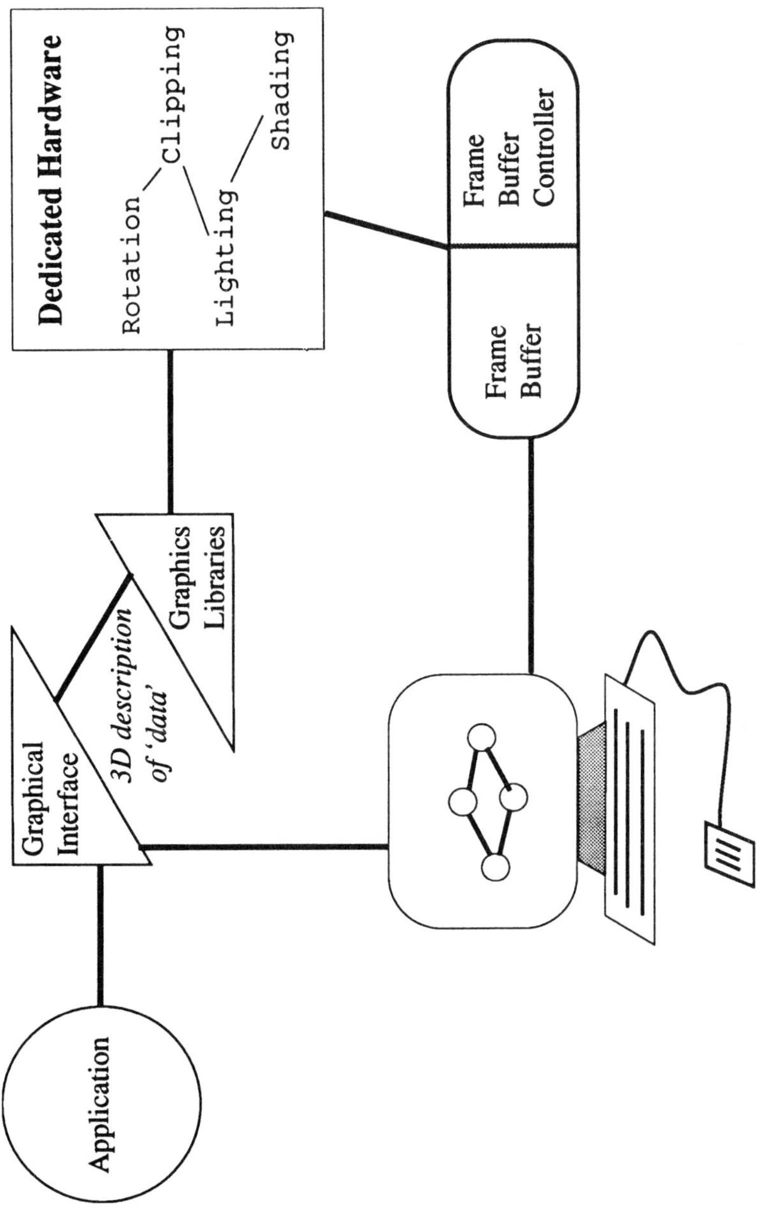

Figure 2.2 An application code uses the graphical interface to gain access to the user and the specialist hardware.

ics libraries also greatly simplifies the task of producing code to generate images of three-dimensional data. One merely needs to provide the three-dimensional description of the system, which often involves describing surfaces as meshes of polygons, and define the viewpoint, lighting and surface characteristics. The hardware performs the rotation, clipping, lighting, shading, and projection operations. It is worthy of note that to view the object from different positions and angles merely requires a redefinition of the viewpoint—it does not involve the tedious exercise of physically rotating the data.

The discussion thus far has dealt with the creation of images on the workstation screen. In the next section I discuss the various ways in which these images may be recorded more permanently.

2.3.2 Obtaining "Hardcopy"

The are many devices available for recording images on paper, film, slides video tape, *etc.* I will distinguish between a device which builds an image from a series of monochrome line segments (*a pen-plotter*, for want of a better name), and one which is able to render raster images.

Of the various rendering techniques that one might use, a few are amenable to description in terms of simple line segments; contour plots and surface meshes are obvious examples. For a purely two-dimensional list of line segments it is a simple exercise to generate the instructions for a pen-plotter which invariably contain as a subset the operations *pen-up*, *pen-down* and *move-pen.* If we decompose a three-dimensional structure in terms of lines then one must perform the three-dimensional to two-dimensional projection of the line list before plotting. The workstation is performing this operation in hardware and the two-dimensional line list that it may be displaying is not directly available.

For images which have been lit and shaded a pen-plotter is not suitable. By far the simplest way of capturing the image currently displayed by the workstation is to take a photograph of the screen. This produces a low-quality image because the screen is a poor quality light source and usually the curved surface of the screen introduces a distortion. An alternative is to use a raster printer which is capable of rendering rows of coloured pixels in a similar manner to the computer screen. Almost all workstation manufacturers provide a mechanism for capturing a list of the displayed pixels on all or part of the screen; a so-called *screen dump.* If one can overcome the language barrier between the computer and the printer then this list of pixels may be plotted. A very common language to use for describing images is PostScript—many printers "understand" this language and for those that do not often one may obtain an interpreter. The most popular modern printer is the "laser" printer which understands only PostScript;

it may be used to produce black and white versions of the colour images produced on a workstation screen.

There are essentially three types of (colour) raster printer which I have found useful.

The **"ink-jet" printer** is a device that simply sprays, paints or dabs different coloured inks to make a pixel map on paper or (sometimes) transparent slides. Output is produced quickly but suffers from a restricted range of colours and low resolution (100 DPI) [11].

Wax transfer printers work by "melting" three sheets of coloured wax, in sequence, onto the paper. The early versions of this printer used the colours cyan, magenta and yellow and produced pixel dots at a resolution of 300 DPI. Unfortunately the transfer mechanism was not able to produce a variation of intensity of colour—a dot was simply there or not. Recently a process know as "dye sublimation" has been developed and each colour may now be applied with a range of intensities but at the lower resolution of 180 DPI.

The film recorder is a device that works by using a fine laser beam to draw directly onto 35mm film. A full colour representation of the image is achieved at 1000 DPI — considerably better resolution than the computer screen; projecting the full 19-inch screen onto a 1.5-inch slide makes some use of this extra resolution. The disadvantages of this device are the initial capital outlay and the time taken to develop the film. The cost of film and developing is rather low.

The final device I would like to mention is the **video recorder**. The computer screen is rather like a television but unfortunately not so alike that one can simply record images with a video cassette recorder. The technical difficulties involved in producing reasonable quality video from an RGB (red-green-blue) channel input to a computer screen are not insuperable and are described in Chapter 3. An alternative is to plug a 'genlock card' directly into the back plane of the computer and tell the graphics hardware to drive it rather than the main screen. Video is not, in general, a high-quality medium as there are only about 512 dots in each 12-inch long row of the screen. The eye is, however, very tolerant of imperfections in a moving image so for three-dimensional images where rotation is important and certainly for animated sequences video is very valuable.

In the next section I outline some of the techniques used to render different types of data.

2.3.3. Rendering Techniques

In general the data that a scientist may wish to draw can be classified in terms of its inherent dimensionality (scalar, vector, *etc.*) and the number of *dimensions* (variables) on which it is defined. A wide variety of rendering techniques are available for presenting data and a brief list of the more common methods (all available within DISPLAY, AVS, Explorer and apE, for example) is provided below. Although most of the above techniques for rendering are well known some may require further explanation.

SCALAR DATA	
Dimensionality	Rendering Method
1	line graph bar chart scatter plot
2	contour pseudo colour plot surface net shaded surface
3	slice to two dimensions iso-value net shaded iso-value surface
VECTOR DATA	
Dimensionality	Rendering Method
2 and 3	“hedgehog” plot

The *pseudo colour plot* is a way of rendering scalar data which splits up the full data range into a series of sub-ranges (usually 256, so that they may addressed with one byte) and then associates a colour with each sub-range. Areas which correspond to each sub-range are then flooded with the appropriate colour. The list of colours used is referred to as a *colour map* and is usually chosen to vary through a smooth sequence of colours or intensities. For example, one might arrange for very negative values to be bright blue, fading to pale blue for small negative values, white for numbers near zero and pale red through bright red for positive values. This non-quantitative method of presenting the data is often a colourful alternative to contour plots. A *surface net* is a very common way to render two-dimensional data. It consists of a perspective view of a connected wireframe of height values which represent the scalar field.

In order to obtain shaded surfaces one must first provide a mathematical definition of the surface and then define its properties. Surfaces may, of course, be defined in many ways but are usually split into a series of contiguous "patches" for the hardware to render. The simplest form of patch is a flat polygon, very often a triangle or square. An very efficient way to define a surface for rendering on high performance hardware is to simply to provide a list of polygons specifying the position and colour of each vertex. The hardware can then light, and project the surface. Surfaces produced in this manner may look angular—having facets like the surface of a jewel. A vast improvement may be obtained by applying a *shading algorithm*. A commonly used technique, called Gouraud shading, is to take the apparent colour of the vertices of each polygon and to smoothly interpolate the colour between them.

If one takes the polygons defined by a surface mesh, defines a colour for each vertex and asks the hardware to shade them a *shaded surface* is produced. Given a further scalar quantity defined on the same grid, the magnitude of the local gradient for instance, the colour at each vertex may be chosen from a colour map and the shaded surface may be smoothly coloured. These concepts are easily generalised for three-dimensional data. An *isovalue net* is produced by linking together points of the same value with line segments. The resulting polygons may be coloured and shaded to produce a *shaded isovalue surface*. Another useful way of viewing three-dimensional data in general is to define a plane which intersects the data and to sample the two-dimensional slice of data cut by the plane. Moving the plane interactively gives a reasonable feel for the nature of the three-dimensional data. For vector quantities the most common form of rendering is a *hedgehog plot* which consists of directed line segments whose orientation reflects the direction of the vector field and length reflects the magnitude.

For rendering the types of objects defined above one must be able to draw (light, shade and project) lines and polygons. Most computer manufacturers quote the rates at which these objects can be generated, "high performance" is at present at or above 5000 lines/second and 100,000 polygons/second. This intense graphical performance allows one to realize the real-time animation of many physical processes. In the next section I will discuss the application of these machines to surface science.

2.4 GRAPHICS IN SURFACE SCIENCE

Often we require a detailed picture of the atomic structure of an interface in order to develop an understanding of its properties. The specification of this atomic structure is, in general, a complicated problem which is usually simplified greatly if the structure of the interface is specified in terms of the bulk structure of the materials present at the interface. If the bulk material is crystalline then its structure is generated by periodically repeating a simple *motif* of atoms all of which are contained in a parallelepiped—the *unit cell.* A structural description of a free surface may thus be obtained by the following conceptual steps. Firstly, we imagine a plane intersecting the bulk material; the orientation and location of the plane is conveniently specified in terms of the bulk unit cell. All atoms to one side of the plane may now be discarded and we may choose a new two-dimensional *surface unit cell* containing a motif of atoms which generate a slab of the surface structure. The resultant *ideal* surface provides us with a reference structure in terms of which the true structure may be specified.

In general the atoms on or near the ideal surface will not be in an energetically favourable configuration and the surface will be unstable with respect to various types of *reconstruction*. Such surface reconstruction usually involves the movement of atoms within a few atomic layers of the ideal surface. In reality, of course, the surface is exposed to attack from various gaseous species and may thus become contaminated with adsorbates. The position of the adsorbates may be specified in terms of an origin located in the plane of the ideal surface. It is not unusual for adsorbates to cause further reconstruction of the surface atomic layers. The process of cleaving a bulk material, reconstruction of surface layers and addition of impurities is often difficult to visualize. When one wishes to determine the structural dependence of material properties, and especially when one is attempting to determine which structures are consistent with various experimental and theoretical information, the interactive graphical display and editing of the surface structure is invaluable.

2.4.1 The SYMM Package

The conceptual route to a surface structure described above represents a steady degradation of symmetry in the system. At surfaces, as in bulk materials, symmetry properties are often vital to understanding physical processes and are also important in reducing the computational costs of calculating properties. The package SYMM [12] is capable of generating a surface by performing the above operations while keeping full account of symmetry. As a three-dimensional structure is cleaved the space group is projected into the appropriate plane group and a two-dimensional unit cell which describes the periodicity of the surface is generated. The surface may be represented as a semi-infinite material or a finite slab. The adsorption of other atomic species may be performed in terms of the fractional coordinate system defined by the surface unit cell or in Cartesian coordinates referred to an origin in the surface plane. Adsorbed species may be automatically symmetry replicated or allowed to break the symmetry.

Currently the SYMM package is implemented as a separate module which communicates the current model of a surface structure to the graphical interface codes after each alteration. The resultant atomistic model may be represented in a variety of ways, including wireframe, CPK spheres, and ball-and-stick graphs. As an example of the use of this package the (100) surface of silicon, with the well known 2×1 reconstruction and a monolayer of absorbed chlorine, is displayed in Figure 2.3.

2.4.2 Properties—Using Existing Applications

The creation and visualization of surface structures is very valuable but it is only the first step towards a graphical environment tuned to the requirements of surface science. Having built an adequate description of the surface the structural information is made available on a *standard format*. The "standard" format used is, of course, non-standard; there is no generally accepted format and the best one can do is to choose a convenient form of data storage and then provide *filters* to and from as many other common formats as possible. For any given application package which feeds off such structural information it is sufficient to provide a filter which will read from the standard format and write into the appropriate format for the application. Once this has been achieved one must inform the interface program of the existence of the application and the interface should then be capable of spawning the application with appropriate input files.

For example, if one is using the DISPLAY package the customization of the environment may be achieved using a simple text file which states the location of the package the names of input and output filters and specifies any any intermediate files which will be used to pass data. DISPLAY then adds the application package to its list of available commands and it appears as a menu item. In this way a whole range of existing surface science application packages are brought within the same environment. For complex calculations which require more sophisticated interaction with the user this may be provided either by the application itself entering into a dialogue or it may be preferable to make the filter "intelligent" so that it can assemble the appropriate information. A clear tendency for future software development is for the graphical user interface to provide a set of simple dialogue "tools" for this purpose. The tools offered within the AVS library have improved dramatically in recent releases.

2.5 FOR EXAMPLE

I have chose two examples to demonstrate the techniques and ideas presented in this article.

2.5.1 Magnesium Chloride—An Industrial Catalyst

In Figure 2.4 a titanium tetrachloride molecule is displayed in close proximity to the (100) surface of magnesium chloride. In order to understand the interaction with the surface the electronic structure of the molecule has been calculated with an approximate description of the surface interaction potential. The initial orientation of the molecule is dominated by the electric field near the surface. In the figure the electric field, (a three-dimensional vector field) is sampled by slicing it with a two-dimensional plane. The values of the vector field at the plane are shown as coloured line segments. The colour of each line segment is related to the magnitude of the electric field at that point. The image was generated on a SGI 4D/220 GTXB workstation using DISPLAY. The plane can be placed an arbitrary position or angle and the vector field is automatically projected onto the plane and the "hedgehog" plot refreshed.

2.5.2 Transition-Metal Surfaces

The complex electronic structure found in transition metals and transition-metal compounds has been the subject of many theoretical and experimental investigations. One of the most detailed probes of the electronic states is obtained by using synchrotron radiation to excite the conduction electrons in copper sufficiently to cause them to exit the surface. Measurements of the energy distribution of emitted electrons as a function of angle (angle-resolved photoemission spectroscopy contain information about the dispersion of the electronic states. In Figure 2.5 the structure of the (100) surface of copper is displayed with the associated photoemission spectra.

ACKNOWLEDGMENTS

I gratefully acknowledge the support of Imperial Chemical Industries.

REFERENCES

1. One of the 12 CCP's involved in coordinating computational science within the United Kingdom.
2. Located at the Rutherford Appleton Laboratory and accessed via a TCP/IP link.
3. DISPLAY has been primarily developed by Dr P. Sherwood.
4. apE Version 2.1 **Software for Visualisation**, The Ohio Supercomputer Graphics Project, The Ohio Supercomputer Centre, Ohio State University, Ohio 431212.
5. **Application Visualization System**, User's Guide 002424-001, Stardent Computer Inc. 95 Wells Avenue, Newton MA 02159.
6. File Transfer Protocol—a widely available direct connect protocol.

7. Network File System—a UNIX-based protocol which allows one to access disks on remote machines in a transparent fashion.
8. Such a harness is under development at Daresbury; the Parallel Library and Network Computing Environment, PARLANCE.
9. AVS and Explorer both contain a communications harness. Also the Network Communication System NCS.
10. A raster is a line of pixels.
11. The resolution of a printer is usually measured in terms of dots per inch—DPI.
12. SYMM was developed by the author from the symmetry analysis within CRYSTAL [13].
13. C. Pisani, R. Dovesi and C. Roetti **Hartree Fock Ab Initio Treatment of Crystalline Systems**, Lecture Notes in Chemistry, **48**, Springer-Verlag
14. M. Kong, **NCS–Network Computing System Reference Manual**, (Englewood Cliffs, NJ, Prentice Hall, 1987).

3

Animation in Surface Science

Ole H Nielsen

Many types of scientific visualization can best be carried out in the form of an animated image sequence which conveys temporal as well as spatial or other information. Animation on workstation-type computers as well as animation using video equipment is described based upon currently available technologies. Since this field is rapidly moving ahead, predictions are made for the development foreseen in the next few years. Although the techniques described are applicable to a wide range of animation problems, special attention is paid to the animation of atoms needed by, among others, solid-state physicists. Examples of solutions developed by us is displayed in the form of images; a videotape could not be included with this volume.

3.1 INTRODUCTION

The complexity of dynamical processes on surfaces and in the bulk of condensed matter is frequently due to the very large number of atoms that participate in the time-development of the system. Therefore, the results of computer simulations on such systems cannot easily be conveyed by a single graph, and the scientist has to employ two- and three-dimensional computer graphics as well as animation in order to understand the phenomena at hand. Many other areas of science face similar complexity, and it is certainly worthwhile to have sophisticated techniques available, such as those offered by contemporary computer- and video-hardware and software.

The present paper describes an approach to the problem of computer animation and video animation, and presents graphics tools that we use.

Some of these tools have been developed at our laboratory, and others are "freeware" obtained via the INTERNET. The aim of our work is to develop animation tools that are interactive and usable in our daily work, and which, in addition, can be used for presentation purposes. Another boundary condition has been the limits of an ordinary research budget. Therefore, we do not attempt to compete with full-fledged computer-visualization centres and their professional-quality, very expensive equipment, and associated long turn-around times.

We describe the process of going from scientific data to animation, which consists of two basic steps: (1) The *rendering* of data from computations or simulations into computer graphics images, and (2) the various methods for converting image sequences into *animation*. This paper relies upon step (1) being well-understood, since there exists a plethora of tools for generating computer graphics. Hence, we briefly describe some of the possibilities that are available today, and subsequently focus upon the techniques that we have found particularly useful in our group's research. Examples of computer graphics images used in surface science illustrate these techniques. The animation step (2) lies in many ways at the edge of current hardware and software technology, and it is all too easy for the scientist (being an amateur in the animation world) to become lost in this jungle. The attempt of the present paper is here twofold:

Firstly, we shall attempt to review at least some of the alternatives currently (end of 1990) possible, and give pointers to developments that can be foreseen in the near future. The perspective given is that of a physicist who applies contemporary video- and computer-equipment for the purposes of scientific animation, and not that of a specialized expert on the topic.

Secondly, we will describe a method that puts animation on every scientist's desktop— and does so today. This is a software solution for a desktop workstation, written by this author, and which has been made available free of cost on the international computer network INTERNET.

3.2 BRIEF REVIEW OF COMPUTER GRAPHICS

Graphics is used by the scientist for communicating information about experimental data and/or theoretical calculations. In its simplest form, the well-known xy plots are used to display function(s) of a single variable such as the incident energy or polar angle. *Terrain plots* employing hidden-surface removal (*i.e.*, you can't see behind opaque objects) as well as *contour plots* are used to display functions of two variables. When showing discrete lattices in two dimensions, such as a spin lattice, checkerboard patterns with some colouring scheme are often used. These time-honoured graphics techniques are still immensely useful when entering the time domain through animation.

The representation of three-dimensional data give rise to a whole new set of problems in computer graphics. Until the arrival of good and inexpensive holographic display devices, we have to deal with projections onto two-dimensional computer screens or sheets of paper. In any case, the image representation of the physical system has to contain sufficient information for the human visual system to recognize it as a *bona fide* three-dimensional object. This important requirement is the basis for an entire industry in three-dimensional computer graphics. For a comprehensive treatment of the fundamentals of this topic see, *e.g.*, the books by Watt [1] and Foley and Van Dam [2]. Here we will merely list some of the important requirements for aiding human vision to interpret two-dimensional displays of data.

Presumably, the most fundamental representation of three-dimensional information is achieved with hidden-surface removal. This point is used successfully in many graphics techniques, and a review of algorithms is found in Ref. [3]. Figure 3.1 exemplifies this technique. The computer implementation of hidden-surface removal is usually done by the *Z-buffer algorithm* [4]. Briefly stated, in this algorithm a data array of z-coordinates (perpendicular to the screen) is maintained for each image *pixel* (picture element), holding the z-coordinate of the object last painted there. New objects replace old ones in the pixel *only if* their z-coordinate indicates that they are closer to the viewer than the object already painted. This algorithm is now implemented in hardware on high-end workstations, and in software in many graphics programs.

In the next step towards realistic three-dimensional graphics, the human visual system derives important information from the *illumination* (or *lighting* or *shading*) of objects [5]: objects brightest on the upper parts appear convex, whereas objects brightest on the lower parts appear concave [see Figure 3.1(b)]. Some simple, approximate methods for lighting three-dimensional objects are *Gouraud shading* [1,6] and *Phong shading* [1,7]. These algorithms are widely used in computer graphics, and are often implemented in computer hardware.

The *stereo technique* dramatically improves three-dimensional perception. Stereo images which correspond to the left and right human eye, respectively, can be viewed directly or through special stereo screens or glasses [8] in order to provide a vivid three-dimensional impression. *Perspective projections* rendering far-away objects smaller than nearby ones yield important visual clues as well.

Colours may be selected to emphasize or de-emphasize certain objects. For example, bright objects stand out relative to dark objects. One application of this idea is *depth-cueing*, where close objects are bright and distant objects become progressively darker according to their distance from the viewer.

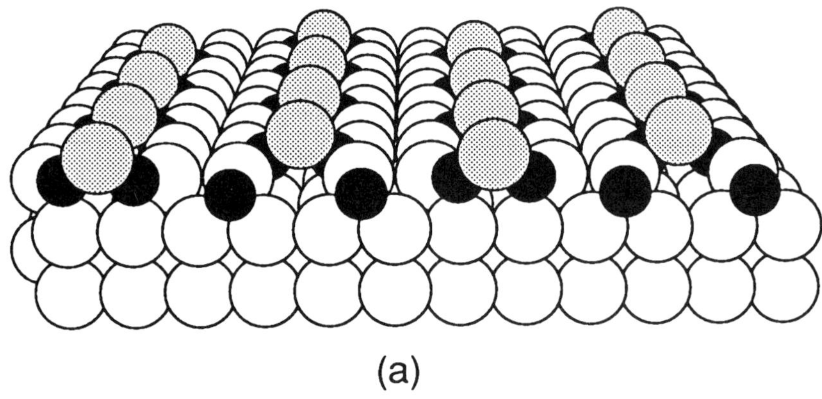

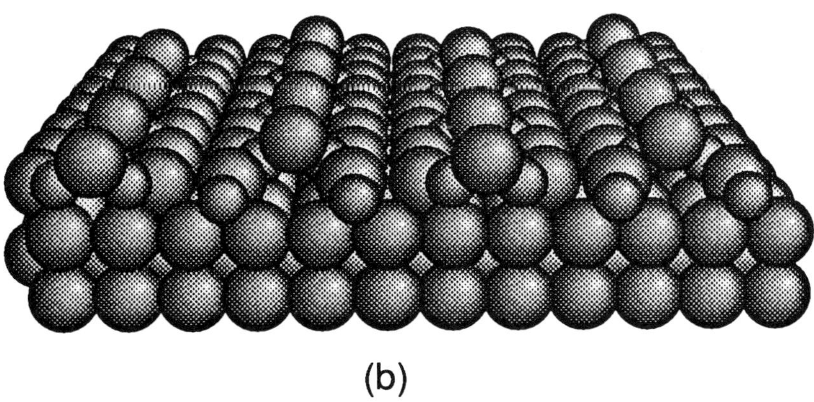

Figure 3.1 Figures of a surface structure with increasing image quality, with atoms represented as (a) plain disks, and (b) monochrome-shaded disks. The same surface structure is shown in (c) with colour-shaded disks and in (d) with ray-traced spheres.

3.2.1 Graphics Libraries for Three Dimensions

As mentioned above, there are very many libraries that will render three-dimensional graphics on a computer screen. The aim is usually to present images as fast as possible, and at the same time try to achieve a reasonable image quality. Some libraries may be specialized for particular applications, others may consist of a set of general tools suited for, *e.g.*, Computer Aided Design (CAD), geographical data, business presentations, *etc.* The prospective user of computer graphics does well in scanning the market for tools that will do his job with as little effort as possible.

In addition to commercial packages, there are almost-free packages developed in governmental or university laboratories. An example of a recent and very general graphics package is Animation Production Environment (apE [9]) from the Ohio Supercomputer Center.

For those who are inclined to program their own computer graphics applications, there are a number of possibilities. In the author's opinion, first priority must be given to graphics libraries that are *standards*, either real standards or *de facto* ones.

For three-dimensional graphics employing the techniques that were discussed above, there is PHIGS (Programmer's Hierarchical Interactive Graphics System) as defined in the ISO 9592 International Standard, and ANSI document X3H3/89-54. Many computer vendors offer PHIGS implementations. In addition there is an on-going effort to incorporate PHIGS into the industry-standard X-window system, denoted PEX (Phigs Extensions to X). PHIGS supplies the elementary graphics primitives such as lines, polylines, polygons, *etc.*, and the PHIGS+ extensions add functionality such as light sources, shading, *etc.*

Other graphics libraries include Silicon Graphics Inc.'s GL, Stardent's Doré, and the GKS Graphical Kernel System ANSI-standard, among others.

3.2.2 Ray Tracing

A widely used technique these days is *ray tracing* [10]. The purpose of ray tracing it is to render images of three-dimensional objects in such detail and with such realism, that ideally one would be unable to distinguish between the computer-generated image and a photograph of real objects. This goal can be achieved more or less with current software techniques. However, the important lesson for the scientist is that it is entirely possible and in fact relatively easy to generate images of three-dimensional objects that "look good," *i.e.*, are fairly realistic approximations to the ideal images that might be desired.

To aid the reader in acquiring the means for ray tracing on a modest PC or a UNIX workstation, we point to the Very Ordinary Rendering

Toolkit (VORT) "free software" which may be obtained from the University of Melbourne [11]. With this ray tracing tool, which we use in our group, the user defines in a simple fashion a collection of objects (such as spheres, cylinders, *etc.*) with geometric properties (such as their sizes and positions in space) as well as material properties (such as colour, refractive index, reflectivity and surface texture). Then light sources are added for illumination of these objects. The "scene" consisting of these objects are then rendered into an image by tracing rays backwards from the viewing rectangle. Here we show a sample input file describing a single sphere, but generally one will produce such files by means of other computer programs. Comments are enclosed in `/* ... */` pairs:

```
/* Define viewpoint and center of view:  */
lookat(100.0, 0.0, 0.0, 0.0, 0.0, 0.0, 0.0)
up(0.0, 1.0, 0.0)                  /* Define "up" direction */
light                              /* Define a light source */
colour 1.0, 1.0, 1.0                         /* White light */
location (200.0, 200.0, 200.0)                  /* Position */
background 1.0, 1.0, 1.0            /* Background color */
sphere                                          /* A sphere */
material 0.0, 0.50, 0.50, 4.0    /* Material properties */
colour 0.000000, 1.000000, 0.000000                /* Color */
reflectance 0.5
radius 60.0
center (0.0, 0.0, 0.0)
```

Processing such a scene file produces a raster image file. For a complex scene, this could take up to a few minutes on a 1 MFLOPS workstation. Such images are useful individually as well as in animation. In Figure 3.1(d) and Figure 3.2 images produced by VORT are shown.

3.3 RENDERING ATOMS FOR SOLID-STATE PHYSICS

Collections of atoms are the basic constituents of the systems that are studied by solid-state physicists, chemists, molecular biologists, metallurgists, and others. Therefore a number of ways have been invented over the years for representing atoms, which of course are quantum mechanical objects that do not have a classical visual representation. We will merely list some of the approaches used on computers, and subsequently describe our approach to the problem.

The classical way of representing collections of atoms is *ball-and-stick models* consisting of wood balls connected by metallic sticks representing covalent bonds, or similar items made of plastic. It is natural to carry this method over to the computer, where one may use balls alone, sticks alone (which are fast to display), or both simultaneously. The rendering of ball-and-stick models may be carried out by the methods described in the previous sections.

Another approach represents atoms as spherical shells or surfaces populated by a thin, uniform distribution of dots. By choosing suitable colours and using depth cueing for the dots, the three-dimensional structure is much more transparent to the beholder.

A different approach emphasizes surfaces of constant charge density, or electrostatic potential, or any other quantity related to the system. The collections of atoms now appear as the equipotential surface of a three-dimensional function, and the surface may be rendered with more or less sophisticated techniques.

Among the very simplest, and consequently fastest, methods we have the technique displayed in Figure 3.1(a). Here atoms are represented as simple disks oriented perpendicular to the paper. The disks are sorted and drawn such that remote disks are drawn first, and closer ones are successively painted as opaque disks of a given colour (here white) with a border (here black) that makes the disks stand out. This algorithm is also known as the *depth-sort algorithm* (Ref. [2] p. 558ff). It resembles the Z-buffer algorithm, but applied to disks rather than pixels. The advantage is that the three-dimensional ball model is achieved with two-dimensional graphics techniques that are much faster than real three-dimensional graphics. It is trivial to draw opaque disks on a printer or a screen using the PostScript graphics language. The following sample code for drawing an opaque circle (the "%"-character starts a comment field) can be printed on any printer or screen supporting PostScript:

```
%!  Draw a full circle at (100,100) with radius 50:
100 100    50    0 360 arc              % 360 degree arc
gsave 0.5 setgray fill grestore         % Fill it with 50% grey
stroke                                  % Draw the circumference
showpage                                % Output the page
```

Of course, one can easily see that such images are not really three-dimensional, but nevertheless they contain ample information for the human visual system to analyze the three-dimensional structure. It is an amusing exercise to produce a stereo pair of images by this method and discover that the image looks like plates of china stacked behind each other (which is really what this technique is all about).

At the next level of sophistication we introduce shading of the disks [5]. If the disks are shown not as disks with a constant colour, but rather as disks coloured as a real sphere would be, a most convincing three-dimensional illusion results. Of course, disks don't overlap properly as real spheres would, but if the system is large enough (more than ten atoms, perhaps), the human visual system will pick up the overall structure rather than being bothered by this inaccuracy.

One may render shaded disks in various ways: The simplest is to draw a large dark circle, which is overwritten by progressively smaller and brighter circles with centres converging towards some point in the upper part of the outermost circle. This technique works very well for producing convincing three-dimensional images on a PostScript laser printer. This is illustrated in Figure 3.1(b).

A more realistic shading appears if one computes the colouring of the disk as if it were a lone three-dimensional sphere illuminated by a light source. A fast iterative method for colouring spheres is the following, whose key lies in performing simple approximations to square roots by a single Newton iteration [12]. Briefly put, for a given point (x, y, z) on a sphere of radius r, and knowing only x, y (pixel coordinates on the screen) as well as r, one must determine the corresponding Z-value as $z = \sqrt{r^2 - x^2 - y^2}$. One usually loops over x values on the horizontal axis, for a fixed y value, with (x, y) staying within the circle of radius r (the boundary of a circle may be computed quickly by the Bresenham algorithm (Ref. [2], p. 441ff). If one uses the value of z_p derived from the previous x-value as an initial guess, the Newton iteration for the new square root z becomes $z = [z_p^2 + (r^2 - x^2 - y^2)]/(2z_p)$. Having calculated the z value, one can very quickly colour a sphere using the Gouraud or Phong techniques discussed above. It is usually advantageous to employ integer arithmetic for speed. Figure 3.1(c) illustrates the result.

The disk approach to plotting three-dimensional structures are used by several contemporary PC graphics packages. At our laboratory has been developed a PC package called **Atomplot** [13], with a substantially faster version for Sun workstations. These packages are extremely valuable tools in our daily work.

3.4 THE ANIMATION PROBLEM

By animation we mean a display of successive images so quick that the human eye perceives smooth motion of the objects in the images. We are all accustomed to animation media in the form of movie reels and video tapes. These forms of animation consist of prerecorded image information usually created by means of a camera recording physical scenes. The animation may be played back using a suitable device (*i.e.*, movie projector or video cassette recorder) in connection with some form of display (*i.e.*, white screen or TV).

When dealing with artificially created image information such as computer images, one cannot point a camera at the scene and record the desired animation. (In fact, one may use a video camera to record events

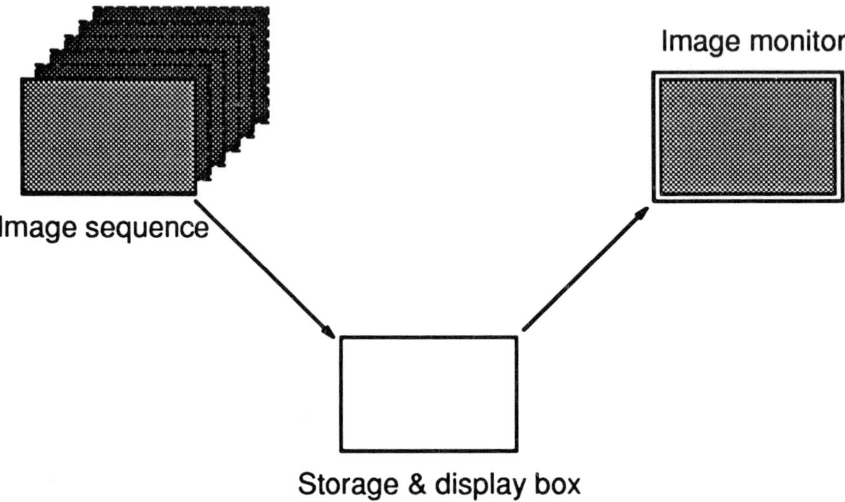

Figure 3.3 Animation steps.

on a computer screen, but due to inconsistent synchronization the result is an intolerable flicker.) In stead, one has to adopt a more generalized view of the animation problem (see Figure 3.3), which we will now discuss in detail.

3.4.1 Image Sequence

The starting point in animation is a source of images in the form of, *e.g.*, film recordings, video camera signal (analog), or digital computer image files. In the case of computer images, they could be generated by any graphics method as discussed in the previous sections, but could equally well be digitized images from other sources such as satellites or TV.

3.4.2 Image Compression

Most types of images contain lots of redundant information, since an image may contain large sub-areas painted by the same, or at least smoothly varying, colour. For example, an image of a square board containing 64 squares that are either black or white could ultimately be described by 64 bits of data. An image of a chess board could thus be described by a particularly simple 64-bit pattern. The purpose of image compression is to invent general algorithms that allow compression with little or no loss of image quality, and a corresponding decompression.

One simple and widely used algorithm is *run-length encoding*, where a sequence of items (bits, bytes, or whatever) that contain repeated instances is compressed by replacing N consecutive instances by some special marker, the number N, and a single instance of the item. Sequences of differing items are left untouched. For images consisting of 8-bit colours (256 colours encoded in 1 byte), it is natural to employ *byte run-length encoding*, as is done on, for example, Sun workstations. We find that this scheme typically reduces image files generated by ray tracing by a factor of 3, whereas simple images may often be compressed 10-15 times. A most important property of this scheme is the simplicity of the compression algorithm, making decompression a very simple and therefore very fast task.

Another widely used compression method is the adaptive Lempel-Ziv coding algorithm [14]. This scheme constructs a table of patterns from the input file, and replaces occurrences of the pattern by a pointer to the table. The look-up table is changed as the compression progresses, adapting to new types of patterns that might occur. The Lempel-Ziv coding is widely used for all sorts of compression tasks, and many UNIX systems offers it as the *compress* command. Public source code for *compress* may be obtained from numerous places. We find that Lempel-Ziv coding generally yields twice as compact compressed image files as with run-length encoding. The reason why we have chosen not to use Lempel-Ziv coding for animation is the complexity of the algorithm, which requires significantly more CPU-power than run-length encoding for a given number of images output per second. An eventual hardware implementation of this algorithm might change this drawback.

Perhaps the most interesting development under way at the moment are the two ISO and CCITT compression standards *J*oint *P*hotographics *E*xperts *G*roup (JPEG) and *M*otion *P*icture *E*xperts *G*roup (MPEG) [15].

The JPEG algorithm, which has been fixed in the fall of 1990, is used to compress still images (*e.g.*, digital photos) by a factor of typically 25 or more while loosing only little image information. It is based upon a Discrete Cosine Transform of 8-by-8 pixel sub-images, after converting the *R*ed-*G*reen-*B*lue (RGB) colour values into luminance (Y) and chrominance (U and V) components. By discarding higher Fourier components and subsequent "thresholding" or discretization of the remaining components, significant compression results. When decompressing, one may choose to render the lower Fourier components first for a rough impression, and subsequently include the higher frequencies for near-perfection. Just as for Lempel-Ziv coding, the JPEG algorithm implemented in software is too slow for animation.

However, new computers are emerging with JPEG hardware built in [15], so we may see a significant increase in performance and utility of

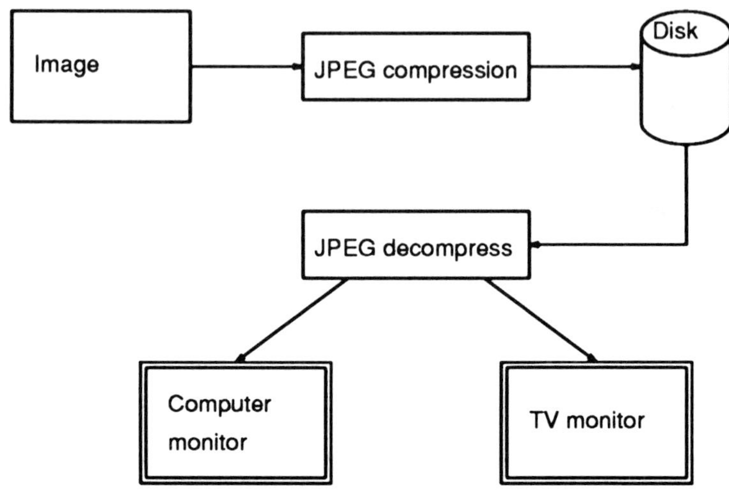

Figure 3.4 Computer system with JPEG compression built in.

JPEG during the next 1-2 years, and it may very likely become useful for animation purposes before long.

The generic way that JPEG compression may be employed in future computers is shown in Figure 3.4: An image is JPEG-compressed, using either software or hardware solutions depending on throughput requirements, before being stored on disk. The disk storage device may consist of any available technology such as magnetic, magneto-optic, WORM, or CD-ROM disks, depending upon how the images are to be used. In order to display the stored images on either the computer monitor or an attached TV screen, a JPEG *hardware decompression module* will be built into the computer. The computer effectively copies the compressed data from disk onto the display device capable of showing JPEG images. The compress-decompress cycle will thus become transparent to the user, and will result in significant savings on storage as well as accompanying increases in effective throughput of image data.

The MPEG algorithm, which is still in an early development stage, is being designed explicitly with video animation in mind. The aim is to display TV-quality video from a slow disk (such as a CD-ROM), requiring a compression ratio well above 100 [15]. At the time of writing the first prototype MPEG decoding chip had been demonstrated by JVC and C-Cube [15]. It is evident that these initiatives by important computer vendors will have a large impact on our way of doing video animation in a few years from now.

3.4.3 Storage and Display Box

This box stores the image sequence in a semi-permanent way, and it has the capability of producing an animated display of the images at the required rate. Examples include: 1) Movie projector for tape reels, 2) standard video cassette recorder, 3) video disks employing media such as magnetic disks, magneto-optic disks, and Compact-Disks (CD), and finally 4) a computer with graphics capabilities.

The central problem is to store vast amounts of data (a digital computer image the size of a PAL-TV image would be as large as 575 by 780 pixels by 2^{24} = 16.7 million colours, or roughly 1.3 Mbytes), and to display them at, for example, 25 images per second. The throughput required might consequently be as high as 33 Mbytes per second. A movie reel easily deals with this complexity because of its two-dimensional storage on a transparent film. A video cassette recorder stores image *frames*† on magnetic tape as consecutive magnetized stripes, slanted at a small angle with respect to the edge of the tape, using helical-scan technology. Video cassette recorders are nowadays simple and inexpensive commodity items.

However, these technologies do not suffice for producing computer animation: Movie reels are expensive and slow to produce, and are very inflexible. Video cassette recorders require an incoming analog video signal at 25 frames per second (PAL-TV), and it is highly nontrivial to have a computer produce images at this rate.

Fortunately there is a way out: Single-frame recording on video storage media. In this technique the computer generates still-pictures on its monitor, one at a time. Firstly, the monitor image must be converted to a standard analog video signal (see *Converting to video signals* below). Secondly, the still-picture video signal must be captured as *a single frame* on the video storage media, *i.e.*, a frame corresponding to 1/25th of a second of video output. At this point in time, the technology for storing single frames is quite expensive (US$ 20.000 would buy a primitive setup). The traditional approach employs sophisticated *editing video tape recorders* writing synchronization marks on the tape, plus an external device which can control and monitor the tape recorder (an *animation controller*). In this technique the tape is rolled back and then accelerated forward (pre-roll), and when the right frame number passes by, the single video frame is recorded on the tape, after which the tape is decelerated. The process may be repeated, recording only 1/25th of a second of animation every half

† A *frame* may be thought of as the fundamental unit of data transmission, whose content can be restored to represent the original image.

minute or so, but it is unfortunately prone to both recording errors as well as mechanical wear.

The animation controller can be eliminated if one is content with only 1 to 2 distinct images per second. In this case, one may succeed in recording, say, 15 consecutive, identical 1/25th-second-frames on an editing video tape recorder. However, this is not a solution for obtaining smooth animation.

Another storage media is disk technology (available from SONY and others), storing single video frames as records on a disk. Afterwards one may select the sequence of video frames that one wishes to display. The great advantage here is a much greater recording speed, as well as negligible mechanical wear on the recording device, plus greater flexibility in accessing frames on the disk. Unfortunately the cost of video recording disks is still much higher than that of video tape recorders, but one may foresee that this technology will become much cheaper in the next few years, considering the boom in Interactive Video applications that is presently starting. The progress in this technology will be accelerated by advances in digital image compression hardware.

With an affordable technology for storing single frame video signals on disks for later display, it is highly probable that a real revolution will occur in the way in which scientists, and everyone else as well, will use the video animation media.

Finally, there is the animation solution which only few people use today: Letting the computer perform animation. Our solution along these lines is described in the section *Desktop animation* below.

3.4.4 Converting to Video Signals

A typical first reaction is "what's the problem" ? Consider, however, that a workstation monitor typically displays 1000 by 1000 pixels on a large screen, scanning the screen line by line 60-80 times a second in order to eliminate perceived flicker. In contrast, a TV screen is theoretically able to display 575 by 780 pixels (PAL-TV), although the resolution in practice is significantly lower. Furthermore, current TV standards specify *interlaced* images for reducing flicker, *i.e.*, every second line of the image is first displayed (in one 1/50th of a second), then the missing lines are displayed (in the next 1/50th of a second). Given that computer monitors operate at frequencies of typically 60-100 MHz and that video signals are typically around 5 MHz, there is a significant down-conversion problem.

The common solution is a *frame scan converter* which stores digitally a copy of the computer's image frames. The converter may be able to adapt to a wide variety of computer monitors. This memory is then scanned at the video frequency, and some form of averaging or interpolation is employed for reducing the resolution to video levels. Current frame scan converters cost typically US$ 15-30.000, available from companies like Lyon-Lamb

[16], RGB Spectrum [17], or Yamashita [18]. Some workstation manufacturers also offer "genlock" or "video" boards (usually for "VME-bus" based computers) for producing a video signal. This may be less costly than a frame scan converter, but it is also non-portable to other computer systems, and there are usually no such boards for low-end desktop workstations.

Another solution is to compromise on the computer's resolution, *i.e.*, stepping down to the typical personal computer (PC) monitor. If the computer is mainly used for producing video images, the resolution of a standard PC display may indeed suffice, and this would be a suitable choice. For the IBM-compatible PCs and Macintoshes there are several relatively inexpensive products on the market today (*e.g.*, Folsom Research Inc. [19], Willow Peripherals [20] or USVideo [21]) which connect to a VGA display adapter and produce good-quality video signals. One should also consider PCs with displays born as TV-compatible, notably the *Commodore Amiga* series.

3.4.5 Image Monitor

This is the device on which the animated images finally appear. The possibilities include white screens, TV monitors, and computer monitors.

3.4.6 Video Animation Laboratory Block Diagram

At the conclusion of this section, we present a block diagram (Figure 3.5) that summarizes the ingredients in a video animation laboratory.

The diagram contains a few alternatives: As the video source one may use either a *frame scan converter* that converts the monitor signals, or (if available) a video board mounted in the computer that is able to output video signals directly.

For the recording method there are three alternatives: Storage on a *video disk*, recording on an *editing video tape recorder* together with an *animation controller*, or plainly recording the video signal as-is. This last solution may be used when the computer is able to provide animation, as described in the following section. A standard video cassette recorder could finally be used for recording on a portable medium. The diagram omits any video editing facilities that might be useful for producing presentation videos.

One must be aware that there exists different TV standards. In most of Europe is used PAL (with a special British variant), France uses SECAM, and USA and Japan use NTSC, all of which are mutually incompatible. You must always check which TV standard is supplied before buying any video equipment. Furthermore, when taking a video cassette to another country, you need to know the TV standard used in that country so that you can arrange for a possible conversion of your cassette. Companies catering to TV-stations usually offer such facilities.

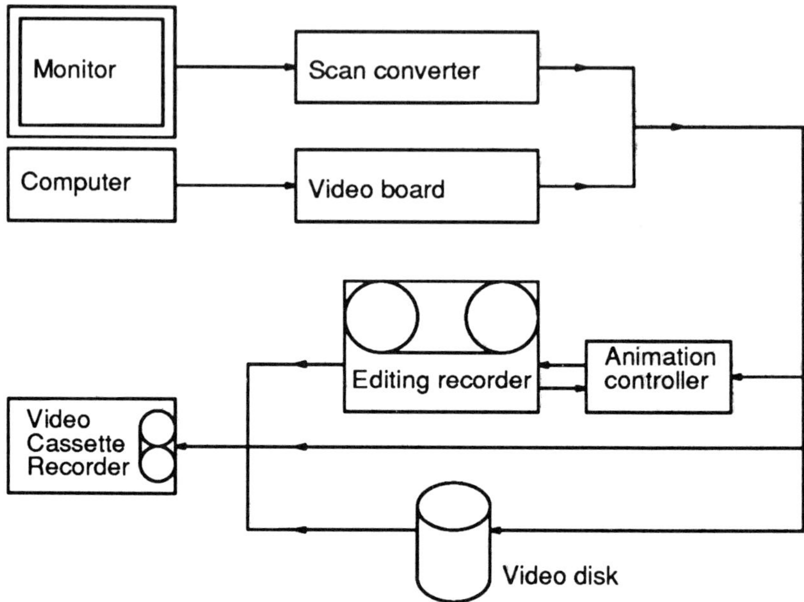

Figure 3.5 Video animation laboratory.

3.5 DESKTOP ANIMATION

The outline presented in the preceding section discussed mainly the "traditional" approach to animation, namely, using sophisticated and expensive video equipment. There exists, however, the alternative of letting a sufficiently powerful computer, such as present-day desktop workstations, perform the animation, and that is the topic of this section.

The most important advantages with *computer animation* are twofold: *economy* and *usefulness*. It is obvious that if you use a computer acquired for other good purposes as a means for generating animation, you have saved a lot of money compared to the large investment in video equipment outlined abovc. The *usefulness* of computer animation as opposed to video animation is significant: every workstation in your installation may be used for animation, meaning that every scientist, student, *etc.* can use animation as a standard tool in their work. At our laboratory, even old, "slow" Motorola-68020 based workstations are being used successfully for monochrome animation, while much more powerful RISC machines are used for the highest performance.

There are virtually unlimited uses for computer animation in scientific investigations, and elsewhere as well. Obvious examples include time-sequences of two- and three-dimensional images, just like regular movies. However, any form of data display depending on time can be animated as well. Furthermore, data depending on *any single parameter* may have that parameter replaced by the time variable in a computer animation. For example, it can be useful to display data for a range of temperatures T (*i.e.*, data from a temperature scan) as animation along the T-axis. This allows the human visual system to pick out changes much more easily than from static data plots.

Video animation still has its place, of course, by as a supplement to computer animation when it is used for presentation purposes, rather than as a laboratory tool. In this case, one by-passes the expensive single-frame recording equipment and records the screen "live." One still has to acquire the means for producing the video signal, however. With such a setup one can easily record animation sequences interspersed with text and still pictures simply by recording onto an inexpensive video cassette recorder. It is even feasible to add sound to each recording sequence. However, more involved editing will require additional video equipment.

A very useful form of computer animation is possible only on high-end workstations with sophisticated graphics-accelerator hardware: if the system to be visualized is sufficiently simple (*e.g.*, a three-dimensional model consisting of a few hundred to several thousand polygons), the hardware may be able to perform the graphics operations at real-time speeds suitable for animation. This solution is less general both because of the high cost of computers with special graphics hardware, and because one loses the real-time capability when the system to be visualized becomes too complex. For example, no current hardware is able to render 10,000 spheres 25 times per second, with the possible exception of a fully configured Connection Machine supercomputer. On the other hand, if the high-end workstation suffices for the problem at hand, one does not need to use any special software for the animation purpose.

At the other extreme of the performance spectrum are the PCs. However, current PCs have very limited memory capabilities and limited internal bus bandwidth as compared to the more expensive present-day workstations. Hence, desktop animation is only possible on a PC for images much smaller than TV resolutions, *e.g.*, at 100×100 to 200×200 pixels. While this may in some cases suffice for producing animation, it is not nearly as useful as techniques that yield full resolution.

In computer animation, the image sequence is stored as a set of ordinary digital computer files. The animation consists of copying image files to the computer monitor so fast that *bona fide* animation is perceived by the viewer. As pointed out previously, one is here faced with a serious bottle-neck, since many Megabytes of image data will have to be copied to the computer monitor every second. In addition, one needs some type of software control over the animation. For this purpose it is reasonable to use today's sophisticated *G*raphical *U*ser *I*nterfaces (GUIs) to create a windowed application with buttons, sliders and such.

An important design criterion for the computer animation software consists of selecting a storage medium for the image data to be copied to the screen. The obvious choice is to copy the data directly from disk storage onto the screen. However, when the animation sequence consists of hundreds of disk files, the operating system overhead involved in opening, copying and closing these files will effectively prevent animation of anything but very small images. Even circumventing this overhead, one must be aware that an affordable, large disk currently available (say, a 1 Gigabyte disk with a synchronous SCSI interface) can deliver a maximum throughput of 1 to 2 Mbytes per second. This limits the possibilities significantly, but with image compression a disk may still be fast enough.

The other alternative is to preload image data from disk into a (large) area of *R*andom-*A*ccess *M*emory (RAM). Here there is no bottle-neck because the RAM access time is typically 80 nanoseconds, compared to 15 milliseconds for a magnetic disk. It is quite realistic to configure a desktop workstation with, say, 32–64 Mbytes of RAM, given the current market price of around US$ 60 per Mbyte of RAM (for a simple so-called "SIMM" memory-module). However, one quickly realizes that even such large memories will only hold rather short animation sequences of TV-sized images.

Whether one uses disk or RAM storage for holding the images, it is obvious that the size of this storage quickly becomes a limiting factor when the desired throughput is several Mbytes per second.

Enter the concept of image compression. If the image data files can be compressed by any significant amount such as a factor of 20, 10, 5 or even 2, it is evident that correspondingly longer animation sequences can be produced with a given piece of hardware. The only condition is that the hardware should be able to decompress the images as quickly as they are needed. On our RISC workstations animation is possible with a speed degradation of about a factor of 2 as compared to animation without compression-decompression.

With these two key technologies, (1) bit-mapped graphics displays on a fast computer, and (2) compression of digital images, it is possible to construct desktop animation software, which is the topic of the next section.

3.6 MOVIETOOL: A DESKTOP ANIMATION TOOL

Movietool is an animation tool based upon the principles discussed in the previous sections. This tool, written by the present author [22], is able to use the *byte run-length encoding* scheme for image decompression, and

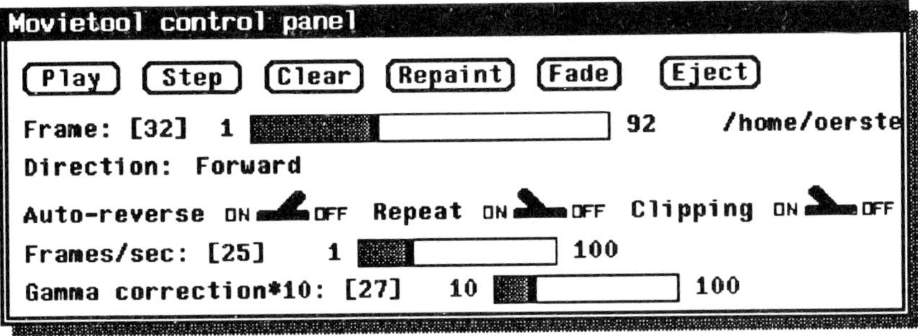

Figure 3.6 Movietool control panel.

the graphical user interface SunView of a Sun workstation (an X-windows version will be available soon). It should be emphasized that the ideas used in Movietool are not at all restricted to this particular brand of computer. It is merely the particular implementation reported here which is programmed using Sun's user-interface, and I recommend implementation on other types of machines using the ideas of Movietool.

Movietool is invoked by the command `movietool [flags] rasterfiles`, where the `flags` are used to customize Movietool for the intended application, and the `rasterfiles` are simply image data files. Having read the files, a *control panel* (Figure 3.6) appears along with a separate *display canvas* containing the animated images. The control panel has been designed starting with the concepts of a video cassette player.

The panel includes facilities for starting and stopping the animation, for controlling direction and repetition of the animation, for positioning at a particular image frame, and for selecting the desired animation speed. The display canvas is a separate window on the screen where the animation can be seen.

Among the customization flags there is the `-encoded` flag, which enables the real-time decompression of run-length encoded images. The `-zoom` flag allows magnification of small images so that they are more easily visible. The `-subregion` flag lets the user pick out part of the image for display. The `-CDROM` flag will pop up a software tool for adding sound from an attached CD-ROM unit to the movie, with the CD player being synchronized to the starting and stopping of the movie.

Any type of image sequences can be viewed using Movietool, be they monochrome or colour (8-bit colour, only, at present) sequences. However, monochrome images are usually 8 times faster to display than identical colour images because there is only 1/8th as much data that must be processed by the computer. For the same reason, a given amount of RAM memory can hold 8 times as many monochrome images as colour ones.

The performance of Movietool in term of frames displayed per second will vary with the computer. For a Sun SPARCstation-1 with a GX graphics accelerator, a full screen (900×1152 pixels) of monochrome images is played at 50 frames per second. Colour images of 500×500 pixels are played at 20 frames per second. If the run-length encoded format is used, the speed typically goes down by a factor of 2.5-3, and less if a faster CPU is used.

These speeds are consistent with a sustained throughput of 5-6 Mbytes per second on the system bus (the SPARCstation S-bus), involving the copying of data from RAM memory to the framebuffer's video memory. This bus is among the fastest in the desktop workstation market, and one would not expect to achieve significantly higher throughput on any desktop computer of the day.

We conclude that Movietool is able to achieve real-time animation of colour images that are only slightly smaller than TV-resolution, but of course with the excellent quality of a digital monitor. It will be necessary to employ image compression on longer animation sequences, but with the most recent workstations approaching the 30 MIPS performance level, this overhead is not that important any more, and it will certainly become negligible within the 1991-1992 time frame.

The use of Movietool for animating surface-science images is reported in great detail in Chapter 4. It is evident from the treatment there that Movietool has lots of possible applications, considering the very many ways that one may analyze data generated by molecular dynamics simulations. For presentation purposes one may often want to use animated colour images, such as the ray-traced ones shown in Figure 3.2. However, monochrome images are often very effective as well, such as the raster-shaded monochrome spheres shown in Figure 5.5 of Chapter 5.

Having seen applications of Movietool from all over the world, it seems to this author that Movietool is genuinely valuable for providing animation on the scientist's desktop at a most affordable price (it is free).

3.7 CONCLUSIONS

Computer graphics is a well-understood technology today, both in terms of software and the required hardware, although one will certainly see dramatic improvements of the state-of-the-art in the coming years. This paper briefly describes some of the fundamental aspects that form the basis for three-dimensional graphics, such as the *Z-buffer algorithm* [4], shading algorithms [1,6,7], stereo and depth-cueing. Current graphics standards such as PHIGS are important to consider, but there are also specialized tools on the market. Ray tracing [10] is the technique of choice for almost-real-looking computer generated pictures.

Solid state physicists and others have a special need for portraying atoms as spheres, and a number of currently used techniques are discussed. A simple technique based upon the PostScript graphics language has been extremely useful in our work, as has an interactive tool called *Atomplot* [13].

An analysis of the concepts in video animation shows that there are a number of solutions, traditionally based upon expensive video equipment. However, one may also use a computer to animate sequences of image files that are stored on a normal disk or CD-ROM disk. In this case, it is important to utilize compression techniques to squeeze large amounts of image data into the available computer storage. This author has developed *Movietool* [22] as a solution based upon a widely used and inexpensive UNIX workstation. Movietool is very useful as a laboratory analysis tool for time-dependent phenomena *etc.*, as well as for video animation in conjunction with video equipment.

Given that there are vast on-going efforts in the industry to bring video animation to the desktop, we are going to see some real breakthroughs in 1991 and in the coming years. Disks for storing video images, both read-write and read-only (CD-ROMs), will proliferate and become much cheaper than what is available today. Image compression techniques such as the JPEG and MPEG standards [15] will become standard tools, with decompression hardware built into disks and even small computers. It is evident that we are just at the beginning of a real revolution in computer- and video-animation !

ACKNOWLEDGMENTS

Financial support from the Danish Research Councils through the *Center for Surface Reactivity* is gratefully acknowledged. Discussions on video animation with the staff at Center for Scientific Computing, Helsinki, and in particular with Otto Pesonen, are gratefully acknowledged.

REFERENCES

1. A. Watt, **Fundamentals of Three-Dimensional Computer Graphics** (Reading, MA, Addison-Wesley, 1989)
2. J.D. Foley and A. Van Dam, **Fundamentals of Interactive Computer Graphics**, (Reading, MA, Addison-Wesley, 1984).
3. I.E. Sutherland, R.F. Sproull and R. Schumacker, "A characterization of ten hidden-surface algorithms," *Computing Surveys* **6**(1), 1 (1974).

4. E. Catmull, in H. Freedman, ed., **Tutorial and selected readings in Interactive computer graphics**, (New York, IEEE, 1980), pp. 309-15. See also Refs. [1], p. 105, and [2], p. 560.
5. V.S. Ramachandran, "Perceiving shape from shading," *Scientific American*, **259**(2), 58 (1988).
6. H. Gouraud, "Illumination for computer generated pictures," *Comm. ACM* **18**(60), 311 (1971).
7. B.-T. Phong and F.C. Crow, "Improved rendition of polygonal models of curved surfaces," in *Proceedings of the 2nd USA-Japan Computer Conference* (1975), pp. 475-80.
8. Simple stereo viewers can be obtained from, *e.g.*, *nu 3-D vu Co.*, 71 East 28th Avenue, Eugene, Oregon 97405, USA.
9. Animation Production Environment (apE), version 2.0, is available from: The Ohio Supercomputer Center, 1224 Kinnear Road, Columbus, Ohio 43212, USA. The electronic mail address is: `apE@apE.osgp.osc.edu`.
10. A.S. Glassner, "Ray tracing for realism," *Byte*, December 1990, pp. 263-71; A. S. Glassner, ed., **An Introduction to Ray Tracing**, (New York, Academic Press, 1989).
11. The VORT raytracing tools are obtainable from: The Software Support Programmer, Department of Engineering Computer Resources, Faculty of Engineering, University of Melbourne, Parkville Vic 3052, Australia. The electronic mail address is: `echidna@munnari.oz`. With access to the TCP/IP INTERNET, VORT can be obtained by "Anonymous FTP" to `gondwana.ecr.mu.oz.au` (128.250.1.63), in the file `/pub/vort.tar.Z`
12. J. Bentley, **More Programming Pearls**, (Reading, MA, Addison-Wesley, 1989), p. 147ff.
13. Please write Dr. Karsten Jacobsen at the same address as this author for the PC version, and the author for the the Sun version.
14. T.A. Welch, "A technique for high performance data compression," *IEEE Computer*, **17**(6), 8 (1984).
15. N. Baran, "Putting the squeeze on graphics," *Byte*, December 1990, pp. 289-94; *ibid.*, June 1991, p. 28.
16. Lyon Lamb Video Animation Systems, Inc., 4531 Empire Avenue, Burbank, CA 91505, USA.
17. RGB Spectrum, 2550 Ninth Street, Berkeley, CA 94710, USA.
18. Yamashita Engineering Manufacture Inc., 559-1, Funako, Atsugi-city, Kanagawa, 243 Japan.
19. Folsom Research Inc., 526 East Bidwell St., Folsom, CA 95630, USA.
20. Willow Peripherals, 190 Willow Avenue, Bronx, NY 10454, USA.
21. USVideo, 62 Southfield Ave., One Stamford Landing, Stamford, CT 06902, USA.

22. Movietool is obtainable from the author's site, using "Anonymous FTP" to the machine **oersted.ltf.dth.dk** (129.142.66.16), in the directory **/pub**, the file names beginning with **movietool***. Outside Europe you should use FTP to **titan.rice.edu** (128.42.1.30). If you don't have access to INTERNET, send electronic mail to the **archive-server@rice.edu**.

4

WINSOM Solid Modeller and its Application to Data Visualization†

M William Ricketts

The IBM UK Scientific Centre's *WIN*chester *SO*lid *M*odelling system (WINSOM) is a set-theoretic, constructive solid geometry (CSG) modeller based on recursive division techniques. It specializes in handling complex models and provides graphical facilities intended for engineering applications. This chapter describes WINSOM and some of the many programs that are linked to it, and gives some examples of their application to problems of data visualization.

4.1 INTRODUCTION

The *WIN*chester *SO*lid *M*odelling system, hereafter referred to as WINSOM, has been under development since 1983 [1]. The main feature that differentiated WINSOM from many other solid modelling systems has been its ability to handle models efficiently. WINSOM is distinguished from other systems in a number of additional respects. First, it is able to represent a wide domain of shapes. A large range of conventional geometric primitives as well as unconventional ones such as numerically-defined fields area available in WINSOM. Shape operations such as bends and blends are also supported.

Second, a number of different methods are used to create WINSOM models. WINSOM has its own simple language, but this has been supplemented by the ESME (*E*xtensible *S*olid *M*odel *E*ditor) language, which has special geometric data types and geometric and programming constructs. The language is designed to support interactive graphical editing. These general-purpose input media are further augmented by a number of programs which generate models from databases, PROLOG fact bases and image analysis.

Third, WINSOM is distinguished from other systems by being oriented more toward picture production than most engineering solid modellers; a wide spectrum of different rendering techniques are available in WINSOM and its associated programs. These range from simple approximate wire frames for quick viewing to ray-traced textured pictures. The range of tasks for which WINSOM has been used is one of the reasons why it has acquired such a wide repertoire of modelling and rendering techniques.

This extends far beyond the usual engineering component design role—one for which WINSOM is perhaps least suited, and which we have hardly explored. The applications that have been explored vary from molecular graphics to computer sculpture and from the generation of single images to the production of ten-minute animations. The only obvious link between these applications is that they all include large sets of three-dimensional geometric data, and WINSOM is thus best represented as a *data visualization* modeller.

4.2 MODELLING TECHNIQUES

While it does not impinge upon the average user, WINSOM is unusual in being a set-theoretic of CSG (constructive solid geometry) modeller. This means that is represents solids as the set-theoretic combination of simple primitive shapes [Figures 4.1 and 4.2]. Thus, shapes such as spheres and cones can be added to or subtracted from one another to create a more complex shape. This contrasts with the majority of today's systems, which represent solids as their boundary surfaces, linked into a coherent whole by topological information. Although many other modellers use the set-theoretic operations as input techniques, they build a boundary model internally; WINSOM does not. Virtues of this approach include numerical stability and compactness; disadvantages include non-uniqueness and the absence of localized surface information, such as edges and vertices.

Historically, the lack of localized information has made computations using set-theoretic models very expensive, since the entire model must be processed to find even the simplest feature, such as part of the surface of the model. WINSOM overcomes this problem by using a technique known as spatial division, which was first implemented in VOLE (*VOL*ume *E*valuator) [2]. Instead of searching the model to try to find its surface, the space that the model occupies is divided into smaller subspaces. For each of these subspaces, new set-theoretic "submodels" are created by replacing those primitives that do not have a surface in the subspace by the empty or universal set and performing elementary simplifications [3]. These new submodels, which are usually smaller and simpler, are valid only in the appropriate subspace.

In the basic WINSOM algorithm for creating pictures, the space occupied by the entire model is a frustum of a pyramid whose apex is at the viewpoint. This is segmented by eight-way division into smaller sections, and this process is repeated recursively. It continues until a subspace is proved to be entirely empty, or until it is the size of a pixel on the screen, meaning that the area of the pixel corresponds to the face of the frustum facing the viewpoint. Thus, the simple submodels are available jut where they are needed to compute parts of the picture.

To enforce hidden-surface elimination, division proceeds in a front-to-back order, and the current picture is maintained using a quad tree. (For a general discussion of quad-tree structures, see Ref. [4]). When a quad has been coloured, the volume behind it is not divided further. By this means, the WINSOM algorithm avoids most of the computation in regions of space that are hidden, even if they contain complex data.

We have improved the original vole algorithm in a number of ways, in particular by dealing more efficiently with the small submodels generated by the spatial division process [5]. This speeds membership testing and also automatically ignores redundant and repeated primitives. It greatly improves WINSOM's performance.

Similar techniques are used for other types of rendering. To produce wire frames, or ray-traced pictures, the division is performed on an original cuboid containing the model and does not terminate until a limiting value of complexity is reached. (This topic is discussed further in Ref. [6]). At this point, FastDraw [7] generates segments of a wire frame; the ray-tracer stores the divided model and then fires rays into it during picture production [8-10].

4.3 INPUT TO WINSOM

As it was first written, WINSOM included a simple input language. This allowed primitives to be defined, transformed, and combined with the set-theoretic operators. Further commands specified the view to be produced and the lighting required. When WINSOM started being used for large-scale molecular modelling, the shortcomings of this interface emerged. It became necessary to generate models automatically. A programmable interface, connected *via* a relational database [11-13] or PROLOG [14,15], extracted molecular information from the Brookhaven data bank [16]. The same interface was also applicable to generating WINSOM models from other databases and PROLOG fact bases.

While the database and image work was very successful, no improvement was made to the *general* usability of WINSOM. In 1986, work started on the ESME language [17]. This language includes conveniences such as special geometric data types (*e.g.*, vectors) and operations (*e.g.*, vector products), together with programming facilities such as variables, loops and subprograms. Other set-theoretic languages such as SML (*S*olid *M*odelling *L*anguage) [18] and SID (*S*et-theoretic *I*nput to *D*ora) [9] have these facilities, but the design of ESME has been taken a stage further by the inclusion of features which are designed to assist the building of interactive graphical editors for particular applications.

Despite the programming excellence of ESME, there are still severe problems that prevent us from writing general graphical editors. These hinge on the non-uniqueness of the set-theoretic model. A single shape may have more than one definition. We have looked at the simplification of set-theoretic models through the elimination of redundant primitives [19] which relieves the designer of some worry about the set-theoretic efficiency of a model.

The set-theoretic definition of a model cannot be deduced from the picture on the screen. Thus, if we try to modify the model by interacting with the picture—for instance, by moving a primitive—the effect depends on the original definition and is thus somewhat unpredictable. One tool that we have found to attack this problem is the active zone concept of Rossignac and Voelcker [20]. The active zone of a primitive is the region of the model that can be affected by changing the geometry of the primitives. It depends on the model definition as well as its shape. By displaying the active zone, we get some indication of what a particular change will achieve [21].

4.4 OUTPUT FROM WINSOM

4.4.1 Static Pictures

The basic WINSOM algorithm produces coloured and shaded images, which may be lit by multiple light sources, but shadows and reflections cannot be modelled. These pictures are more than adequate for most of the data-visualization applications with which we are involved. As mentioned above, there is also a ray-tracing module in WINSOM and this is able to produce the shadows and reflections so beloved by computer graphics aficionados. There is a time penalty for this, although in this program the spatially-divided model may be re-used for a number of views. At the other extreme, low-resolution pictures are easy to produce quickly using the WINSOM algorithm and are widely used for model and view verification. Wire frames may be generated using the FastDraw program, and these can be animated on a device such as the IBM 6090. Figures 4.3–4.5 show the different types of rendering produced by WINSOM: Figure 4.3 shows the central WINSOM algorithm; Figure 4.4 shows ray-tracing into a divided model; and Figure 4.5 shows wire-frame generation. This model is made up of 55 planes, 131 spheres, and 340 cylinders. Wire-frame output is about four orders of magnitude cheaper than ray-tracing, with the central WINSOM algorithm positioned between these two extremes. However, these differences are reduced for complex models.

FastDraw sacrifices accuracy to speed. As division progresses, the ratio of model complexity to subspace size is monitored. If it becomes too large, work on that subspace is halted and the submodel is abandoned. This avoids spending time on complex regions that contribute little to the visual impact of the final model. For models containing hundreds of primitives, these wire frames can be computed in under a second on an IBM 3081 processor and are used for verifying input and some view planning. The wires in the wire frame inherit the colours of neighbouring surfaces to enhance the legibility of the display. Nevertheless, as with all wire frames, interpreting the picture becomes difficult as model size increases. FastDraw can also prepare a polygon surface from a model. Using polygon sorting or z-buffering (a technique for the removal of hidden surfaces by discarding all but the nearest surface at every pixel) on appropriate hardware, this can give a series of low-quality rendered hidden-surface views in close to real time.

4.4.2 Animation

Static WINSOM images can help the researcher to visualize, analyse, and validate data and support proposed theories. Over the last few years, we have increasingly been drawn into using WINSOM to produce frames for (longer and longer) animation sequences, and we are demonstrating that animation can dramatically enhance visualization. Not only can animation provide structured "tours" in, around, and through our solid models, but it can also convey the evolution of those models in time—a capability which can be of great value in scientific applications.

4.4.3 Nongraphical Output

WINSOM is able to compute volume and surface area properties (which traditionally are used to prove that the model really is a solid). The ability to obtain nongraphical data interactively is more interesting: a tablet is used to identify a point on a picture. Once the visible object is found, properties such as surface normal (*i.e.*, surface orientation, colour, and location can be identified within the model definition [22,23].

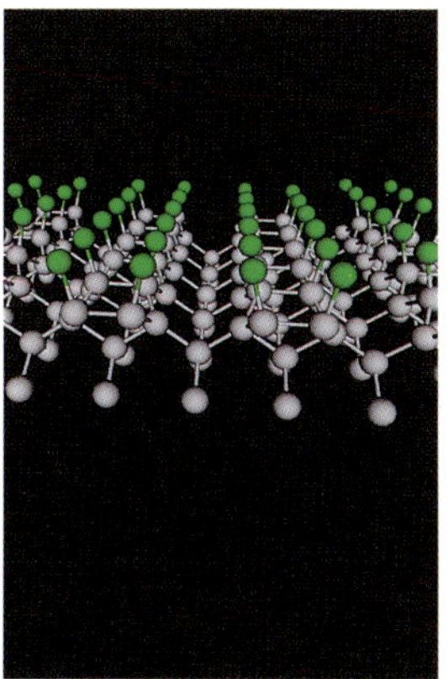

Figure 2.3 A 2×1 reconstruction of the silicon (001) surface with an adsorbed monolayer of chlorine. This figure and Figures 2.4 and 2.5 were produced by DISPLAY 5.2.

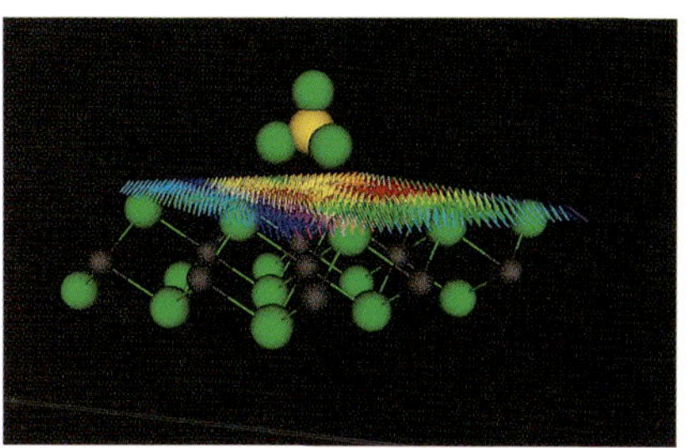

Figure 2.4 A titanium tetrachloride ($TiCl_4$) molecule in close proximity to the (100) surface of $MgCl_2$. The local electric field is shown as a coloured "hedgehog" plot.

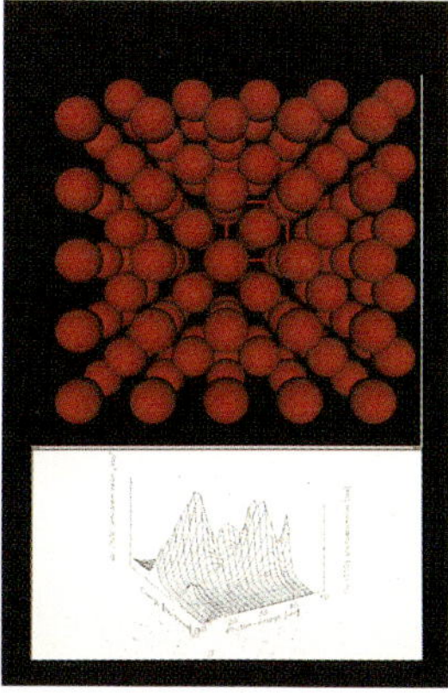

Figure 2.5 The (001) surface of face-centred-cubic copper with the associated photoemision spectra.

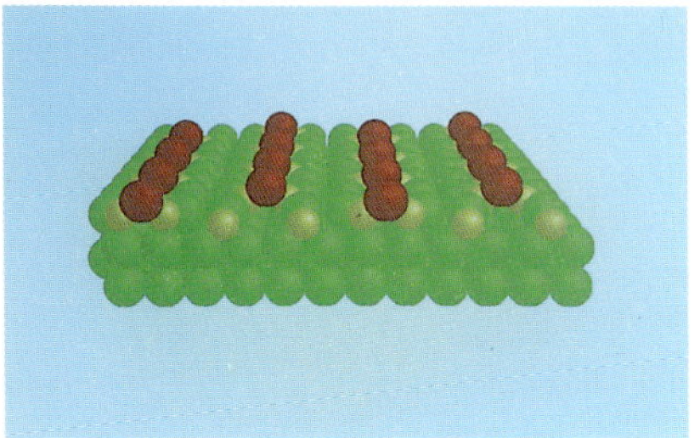

Figure 3.1 The same surface structure as in Figure 1(a) generated with (c) colour-shaded disks, and (d) ray-traced spheres.

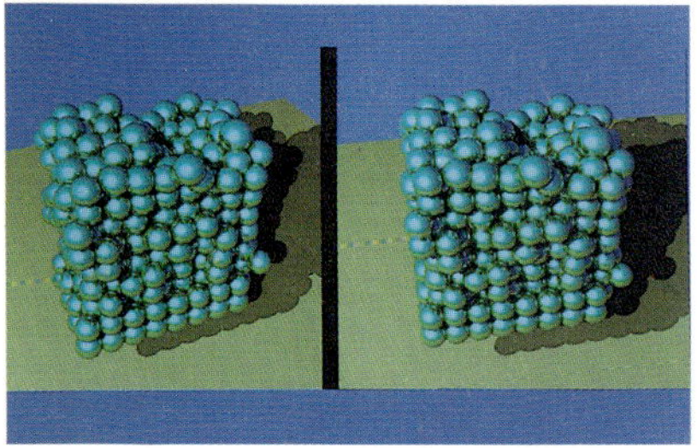

Figure 3.2 Stereo images of two sintering clusters of copper atoms (a), and an aluminium surface slab (b).

Figure 4.1 The geometric primitives available in WINSOM.

Figure 4.2 Operations in WINSOM. Beginning at the top and moving clockwise, examples are the union, intersection, and difference.

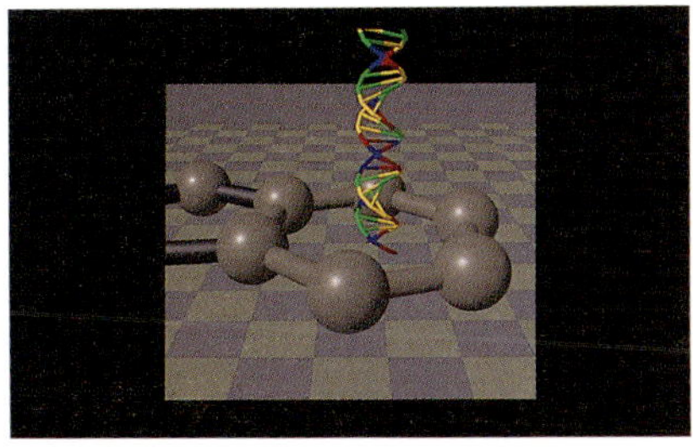

Figure 4.3 An example of the central WINSOM algorithm.

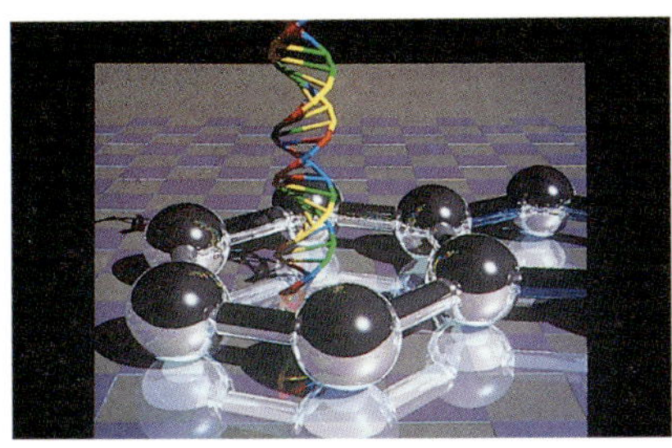

Figure 4.4 A Ray-traced rendering of Figure 4.3.

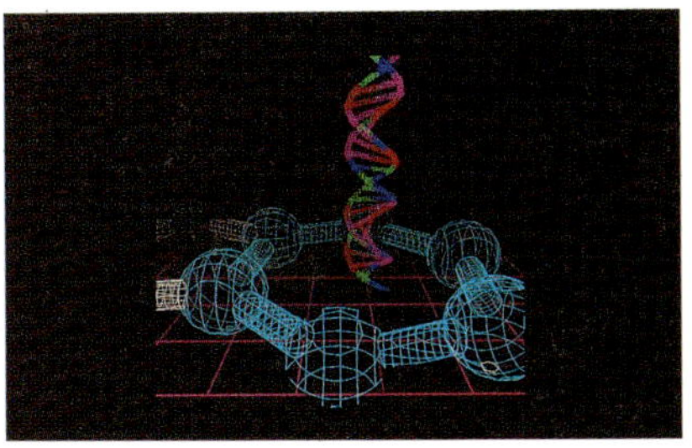

Figure 4.5 A Wire-frame rendering of Figure 4.3.

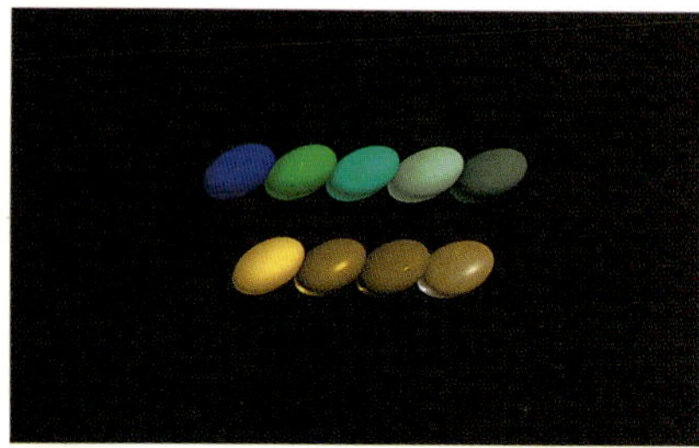

Figure 4.6 Colour and highlighting of surfaces in WINSOM.

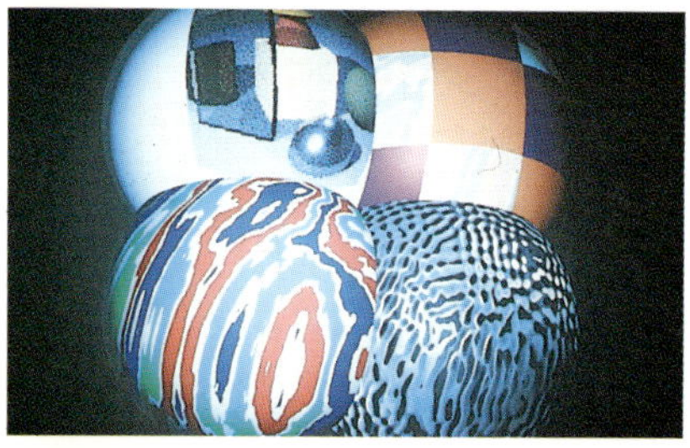

Figure 4.7 Solid textures used by WINSOM.

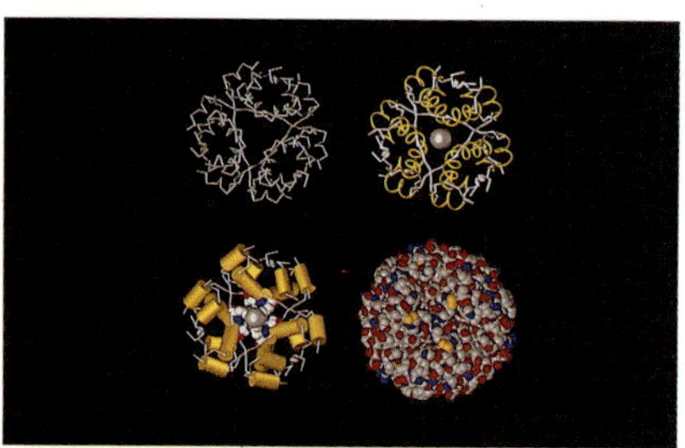

Figure 4.8 Four views of the insulin hexamer.

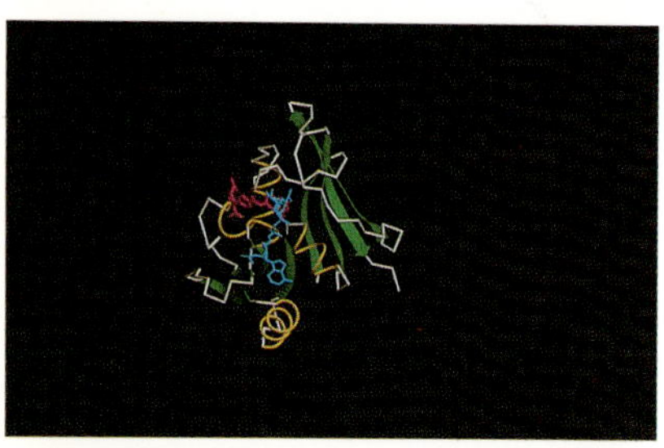

Figure 4.9 Stick-and-sheet models of Dihydrofolate Reductase (DHFR) from the bacterium Lactobacillus Casel.

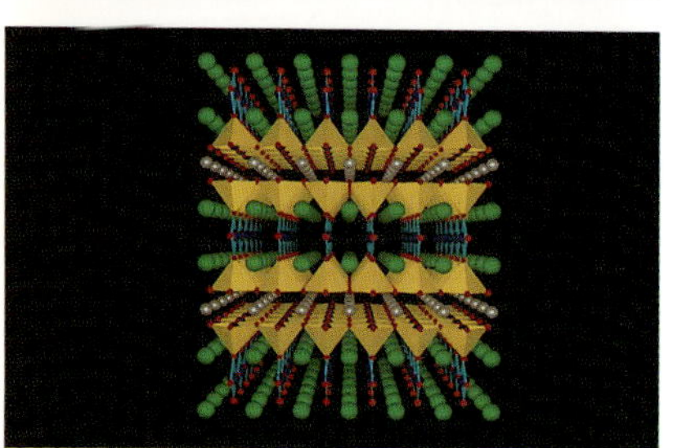

Figure 4.10 Crystal structure of the high-temperature superconductor yttrium barium copper oxide.

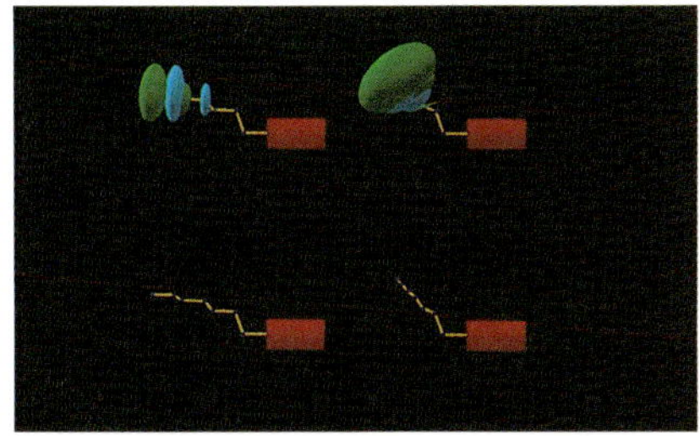

Figure 4.11 Representations of theoretical predictions of average shapes in flexible liquid crystals.

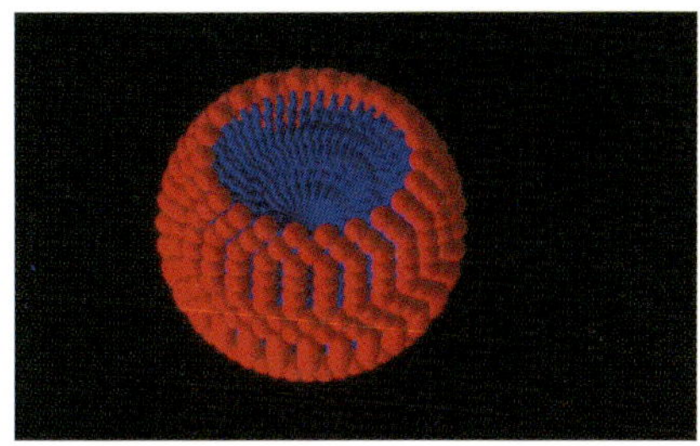

Figure 4.12 Idealized representation of the spherical structure of a micelle.

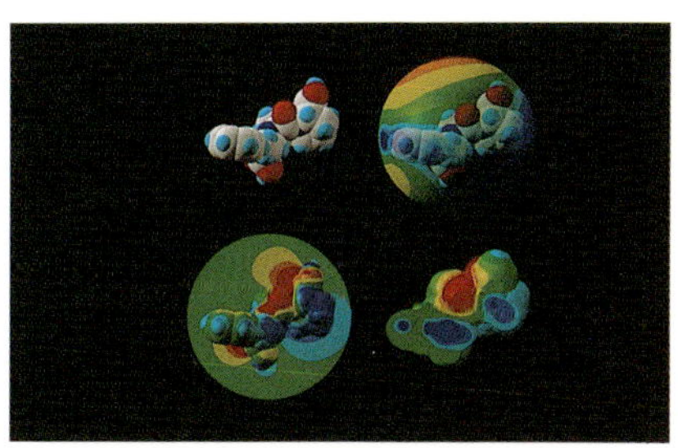

Figure 4.13 Representations of electric field strength outside an enapryl molecule.

Figure 4.14 A series of equipotential contours surrounding an enapryl molecule.

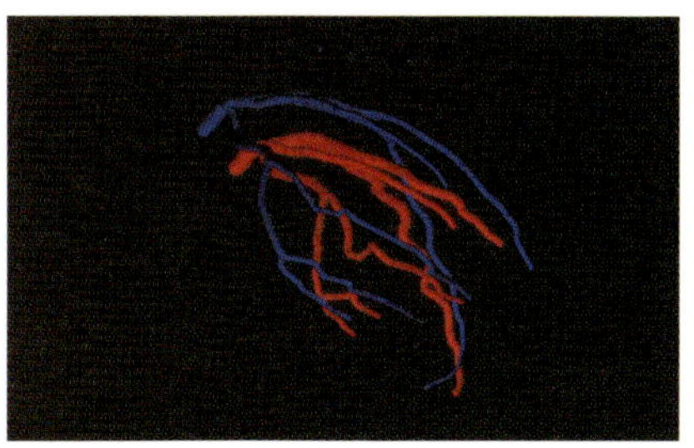

Figure 4.15 Coronary arteries modelled from X-ray data. This figure was constructed from the union of 395 cylinders and spheres.

Figure 4.16 A hippocampal neuron, generated by the union of 1149 spheres and 1134 frusta of cones.

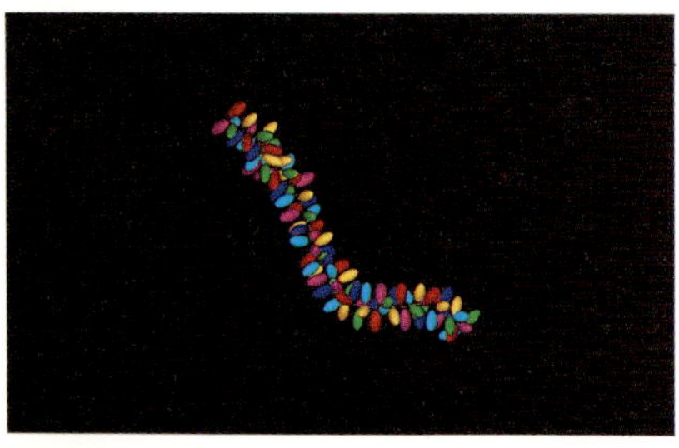

Figure 4.17 One possible configuration of a proteo-glycan assembly.

Figure 4.18 Example of simple primitives blending to give a complex shape, showing a Greek urn (top panels) constructed by blending a crude initial model (bottom panels).

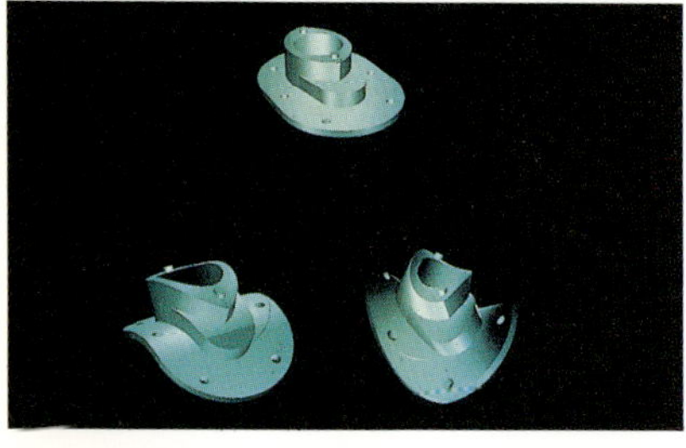

Figure 4.19 The bending and twisting of a set-theoretic model, showing a pump cover plate (top), which is first bent (bottom left), and then twisted (bottom right).

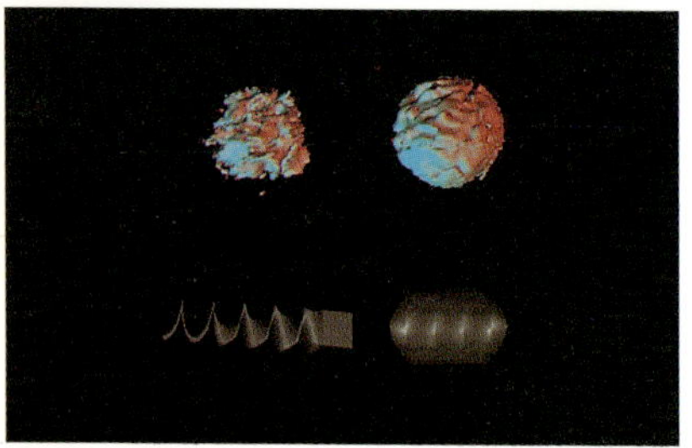

Figure 4.20 Continuous deformation between two set-theoretic models.

Plate 7

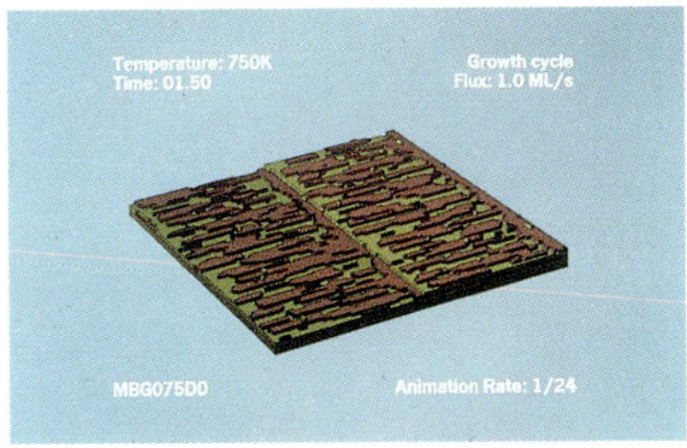

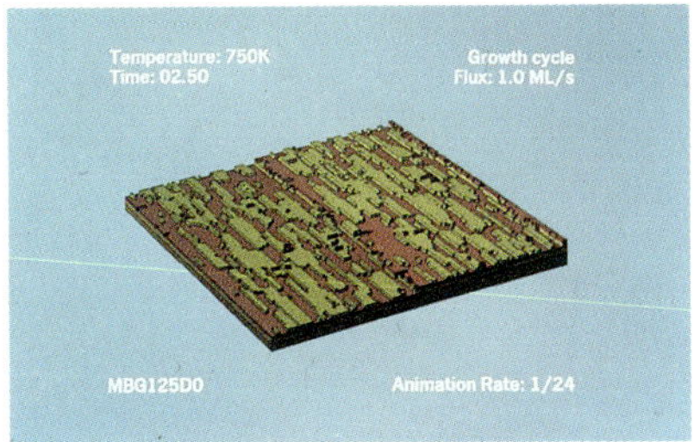

Figure 6.4 The morphology of a substrate with two biatomic steps after the deposition of (a) 1.5 and (b) 2.5 monolayers of material at a substrate temperature of 750 K and with a flux of 1 monolayer/s. The dimension of the substrate is 120×120 lattice sites, and the two colours represent the two domains, with the type-B terraces shown red, and the type-A terraces shown in yellow.

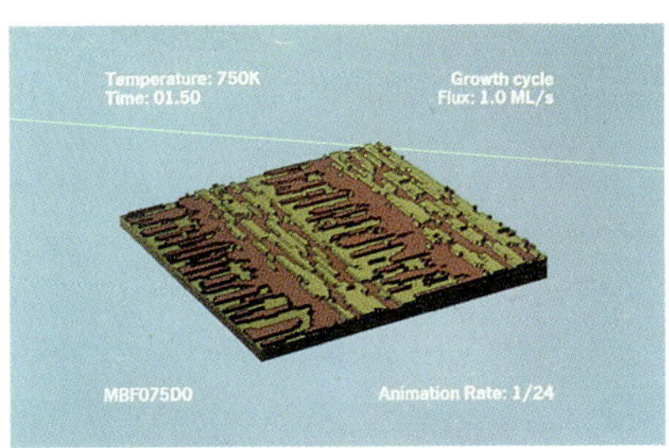

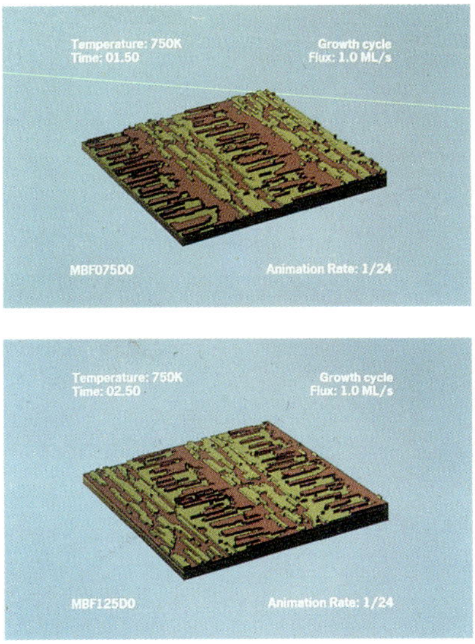

Figure 6.5 The same views as in Figure 6.4, but for a surface with four monatomic steps.

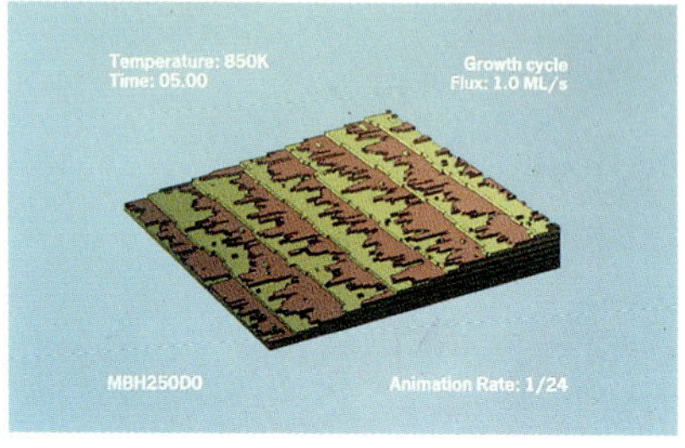

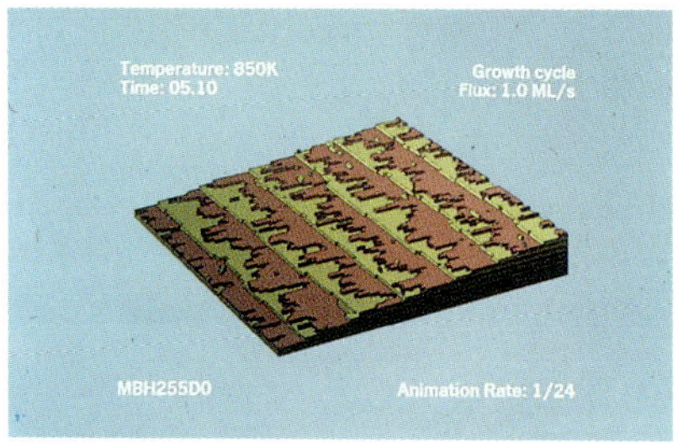

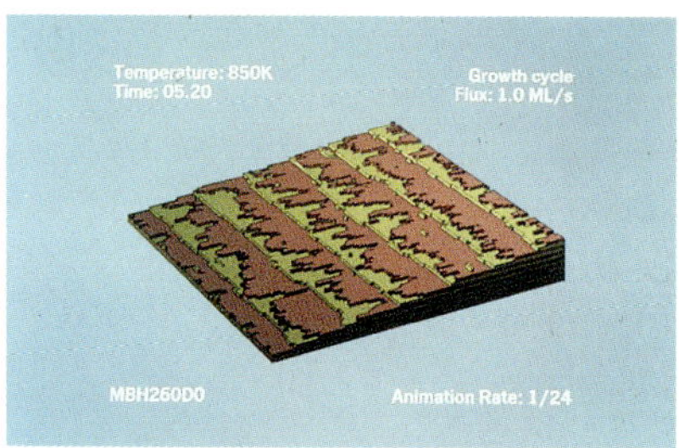

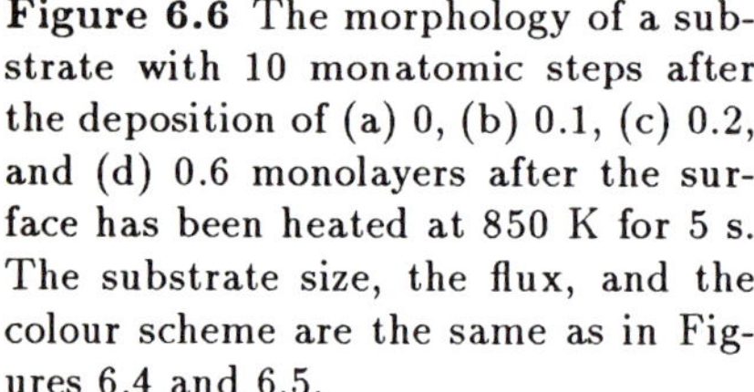
Figure 6.6 The morphology of a substrate with 10 monatomic steps after the deposition of (a) 0, (b) 0.1, (c) 0.2, and (d) 0.6 monolayers after the surface has been heated at 850 K for 5 s. The substrate size, the flux, and the colour scheme are the same as in Figures 6.4 and 6.5.

4.4.4 Low-level Detailing

Within the WINSOM environment, there are a number of utilities which, although not associated with modelling, contribute strongly to the system's ability to produce acceptable pictures. One problem encountered early in producing coloured images is that of obtaining shades of colour that have the user's approval. We have extended the Colour Naming System [24] to include names for surface rendering parameters such as glossiness and polish [25] (Figure 4.6).

The following commands were used to generate the picture in Figure 4.6:

```
DRAW o at(1,9,0) cnsx(blue)
   U o at(3,9,0) cnsx(green)
   U o at(5,9,0) cnsx(bluish green)
   U o at(7,9,0) cnsx(grayish bluish green)
   U o at(9,9,0) cnsx(medium grayish bluish green)

   U o at(2,6,0) cnsx(yellow)
   U o at(4,6,0) cnsx(glossy yellow)
   U o at(6,6,0) cnsx(very polished glossy yellow)
   U o at(8,6,0) cnsx(plastic matt yellow)
```

The first row in the figure shows control of basic colour. Blue, green, and bluish green control hue, grayish controls saturation, and medium controls brightness. The second row in the figure shows control of the highlights (specular reflection). Glossy and matt control the strength of highlight, very polished controls the size of the highlight, and plastic controls its colour.

Having dealt with colour, we must determine the appropriate lighting for a scene. Positioning and directing lights in the imaginary space of a solid model is known to be difficult. We believe that our SLED (*S*olid *L*ighting *ED*itor) program [26] presents an original solution to this difficulty. WINSOM is able to produce an augmented image type, in which pixels are labelled with depth and surface normal as well as colour. sled accepts a picture in this form and allows the user to vary the lighting parameters. Because a normal WINSOM run does not produce shadows and reflections, the result of new lighting parameters can be displayed explicitly. Working at low resolution, the system is fast enough to give an interactive "feel."

However, the main feature of sled is that the lights can be specified both by their position (which can be difficult to fix) and by their effect on the scene. The user specifies the highlights on the displayed object, and the system works backward from these specifications to determine the light position. The angle of the beam and its position can be determined. This is important because the user can indicate the size as well as the position of the desired highlights.

WINSOM also includes texturing facilities [25]. This uses "solid texture" [27] which is shaped with pseudo-fractal or band-processed noise, generated with a minimum of computation from interfering random waveforms, or with regular patterns, such as three-dimensional chess boards. The texture generates values on the surface of an object, which can control all the surface parameters such as hue, gloss and transparency (Figure 4.7). Achieving the desired texture is more difficult than determining a colour. We have not found a convincing English-like description of texture and are experimenting with a texture editor (ted) giving real-time control of parameters by dials or a mouse. ted will combine interactive colour definition with the ability to play with texture parameters such as fractal dimension.

Finally, we have used error diffusion [28] which exploits shortcomings in the human visual system, to produce pictures on a relatively low-colour-resolution display that are in practice indistinguishable from the high-colour-resolution equivalents. Error diffusion is especially difficult when synthetic images are displayed, because they may have large areas of perfect surface which are not found in real images. Important features of our error-diffusion process [29] are the careful choice of colour levels and the use of smooth-area detection and a randomized Peano diffusion path to prevent detectable error-diffusion pattern regimes. All the pictures shown have been error-diffused with a palette of only 252 colours (7 levels—including zero—of green, and 6 levels each of red and blue).

4.5 APPLICATIONS

4.5.1 Molecular Modelling

The first serious use of WINSOM was to model molecules. There are many programs which can generate ball-and-stick representations of molecules, and this simple geometry can be displayed by special algorithms or in real time on vector displays. Our programs [30] are more flexible, since they use the database to collect and structure the data required for a picture, resulting in the generation of an appropriately coloured mixture of atom, ball-and-stick, and backbone representations.

WINSOM primitives for cylinder and helix are particularly useful for an abstraction of molecules to be drawn showing the secondary structure. The application programs require some geometry code to fit these primitives to sets of atoms, when a wide range of representations of molecules becomes possible. In Figure 4.8(a), the representation traces the course of the polypeptide chain through each of the six monomeric units, as white folded "string." The second representation [Figure 4.8(b)] highlights areas where the polypeptide chain folds into regular, stable, helical structures known as alpha-helices. The alpha-helices are shown in yellow spirals. The central zinc atom is shown as a sphere. In the third image [Figure 4.8(c)], alpha-helices are shown as yellow cylinders. Spheres are again used for the zinc atom and the atoms directly involved in bonding between the insulin hexamer and the coordinating central zinc atom. Finally [Figure 4.8(d)], we see an entirely space-filling model, with electron clouds represented by spheres of prescribed radii. The spheres are colour-coded to distinguish different atoms.

All our programs that generate WINSOM models from a molecular database have the option of generating a simple wire-frame equivalent first (Figure 4.9). This permits the user to make a quick check of the picture to see that it has the expected content, and the user can choose a view interactively [31].

4.5.2 Crystals

All crystal structures—from elements in the solid state, through simple metallic salts, to complex minerals—are uniquely described by the positions of the constituent atoms in the unit cell. The unit cell can be thought of as the smallest building block of atoms which, when tessellated indefinitely in three dimensions, yields the repetitive, infinite structures characteristic of crystals.

If the unit cell is described using WINSOM primitives (spheres for atoms and planes for anion coordination polyhedra around cations), then the large-scale repetitive crystal structure can easily be generated. This approach has been used to great effect in the illustration of the structural relationships between some of the recently discovered high-temperature superconductors. The effect of chains and sheets of copper atoms on T_c (the temperature below which the superconducting behaviour appears) can be demonstrated. Figure 4.10 shows a high-temperature superconductor (yttrium barium copper oxide).

4.5.3 Liquid Crystals

Flexible liquid crystals are liquid crystals of the kind found in electronic displays and are composed of elongated particles with a rigid core section. Flexible hydrocarbon chains attached to the rigid core of the molecule enhance its liquid crystalline properties. In Figure 4.11 we show the results of theoretical calculations on these chains. The rigid core is represented by the red cylinders, while the average position of each of the carbon atoms in the chain is represented by the small blue spheres joined by yellow cylinders. The ellipsoids in the top row represent the extent of thermal fluctuation of the atomic positions away from their average location.

Lyotropic liquid crystals are surfactants which include detergents and are composed of molecules with a water-attracting or hydrophilic group and a water-repelling or hydrophobic group. The hydrophilic group is happy to be surrounded by water molecules, the hydrophobic group is not. As a direct consequence, the surfactant molecules tend to arrange in aggregates such that the hydrophobic groups are protected from the water by a shell of hydrophilic groups.

These aggregates of molecules are called micelles. The simplest shape they take up is a sphere. Ellipsoidal, cylindrical and bilayer structures are also known. In living organisms, cell membranes are known to be composed of a bilayer of surfactants; Figure 4.12 shows an idealized representation of these aggregates. We have represented the hydrophilic part of the molecule as a red sphere and the hydrophobic part as a blue worm-like chain.

4.5.4 Potential Fields

Shape is not the only consideration with molecules, which are surrounded by a cloud of electrons that govern interaction with other molecules. In applications such as the study of the chemistry of biological processes, and in the computer-aided design of new drugs, the electronic shape is as important as the conventional shape.

WINSOM has facilities to display potential and other scalar fields from a regular grid of sample measurements over the region of interest. There are two basic facilities provided, allowing either the value or the shape of the field to be displayed [32]. The simpler option is probably to display field values using colour. WINSOM allows any surface to be coloured from the value of a field. This technique is well-suited to cases when the field value is of most interest on a particular surface (as, for example, a molecular surface). However, it is not restricted to this case and can be used to show the values of a field either within an object or in the space around it. In the first case, the object must be sectioned, which is of course easy to do in solid modelling. To display the field outside an object, a surface must be extended into the region of interest. In Figure 4.13, the bottom left panel shows an enalpryl molecule coloured by field strength, with a disk through the molecule, as the added surface.

To provide for the display of field shape, WINSOM has a primitive which is defined by a three-dimensional contour of a field. Such equipotential primitives act in exactly the same way as geometrically defined objects; they can participate in set-theoretic expressions; their surface colour and characteristics can be defined, and they are rendered accordingly.

The bottom right panel in Figure 4.14 shows a series of equipotential contours surrounding an enalapryl molecule. The inner ones are exposed by subtracting a sphere.

4.5.5 Biological Applications

If a pair of X-ray pictures is taken from different angles, and one can find corresponding details in each view, it is possible to reconstruct the exact three-dimensional position of these details. In X-ray pictures of the chest, one can identify details on the arteries surrounding the heart—in particular the positions where they divide—and thus determine where arterial bifurcations are as three-dimensional positions within the living body. One can also estimate the widths of the arteries and so build a model of the arterial system surrounding the heart. This is most simply represented by a number of connected cylinders and cones. Repeating this with each frame from a cinematic X-ray film produces an animated model of the arteries in action during the heartbeat. Figure 4.15 shows two frames from such an animation overlayed, showing the left ventricle in its most contracted state (in red) and in its largest state (in blue). This method offers considerable potential for the diagnosis of heart diseases.

In a project concerned with gaining a greater understanding of the operation of the Hippocampus region of the rat's brain [35], WINSOM is used to display a solid model of a single neuron. Photographs of optical sections of the entire cell volume, obtained by utilizing the small depth of field of a microscope, are image processed to find the centre line and width of the "in focus" parts of the dendrite structure and its topography in three dimensions [36]. This geometric information is used to construct a WINSOM model as a union of spheres and frusta of cones. The picture shows one view of the three-dimensional dendritic structure of a typical cell. The measured volume and surface area of the structure are used in theoretical predictions, which are then compared with electrophysiological measurements on each cell, with the aim of broadening the theory of cell operation (Figure 4.16).

Another biological application is the study of proteo-glycan assemblies. These are very large molecules whose approximate chemistry is known, but the only evidence for their shape comes from electron micrographs. We have used the programming power of ESME to produce a range of possible shapes based on available statistical information. One configuration of proteo-glycan assembly is shown in Figure 4.17. Probably none of these is a "correct" picture of a proteo-glycan assembly, but each gave the researchers a much clearer idea of the three-dimensional nature of the molecule.

4.5.6 Physics

Recent work on the simulation of molecular-beam epitaxy illustrates how theory can explain the growth of semiconductor surfaces under the influence of directed atomic and molecular beams. The theory was known to be consistent with gross experimental observations, but a WINSOM animation (see the article by Wilby and Clarke in this volume) of the time-dependent processes yielded much information on the nature of individual growth mechanisms, which was fundamental to the understanding of the epitaxial process.

Another scientific animation addresses even more fundamental issues—those of quantum chromodynamics, a discipline that attempts to comprehend the basic forces of nature. Earlier work in this area [37] has used WINSOM to visualize the energy fields surrounding t'Hooft-Polyakov magnetic monopoles, a concept predicted by recent theories. The animation combines this work with results stemming from a branch of mathematics known as topology to show how the shape, motion, and even the basic nature of a monopole can change in the presence of another monopole.

4.6 FURTHER DEVELOPMENTS OF WINSOM

Much effort has been put into helping potential users of WINSOM get the best from their system. This time has rarely been wasted, as applications often lead to important developments. In addition to such fortuitous advances, we are currently focussing on a number of facets of the system.

4.6.1 WINSOM II

Various special applications of WINSOM caused us to extend the conventional set-theoretic modelling capabilities in new directions. We now realize that many of these extensions can be combined under a single umbrella of algebraic modelling. WINSOM II is an experimental system that implements some of these ideas.

Attempts to use WINSOM in design (as opposed to science and engineering) led to the implementation of a blend system similar to that of Ricci [39] (Figure 4.18), and thus to several operations generalized beyond the usual set operations. Design applications also required perspective and then more complex viewing projections. We found these could be used to distort the objects themselves, as well as our views of them. This permitted blending and bending operations on solid objects (Figure 4.19) similar to those of Barr [40].

The introduction of contoured potential fields [32] described elsewhere in this paper (see also Figure 4.14) was the key to the integration of these concepts. In WINSOM II, all objects are considered as fields. No longer does a primitive sphere just have an inside and an outside, but rather every point in space has a potential that gives some measure of how far inside or outside the sphere it is. To make a solid, we contour the potential: points with a negative potential are inside the sphere, and points with a positive potential are outside the sphere.

The operations of union and intersection of solids are replaced by the minimum and maximum of potentials. As a potential defines a number at each point in space, we can also apply the standard arithmetic operations to add and subtract potentials, or even to take sines and logarithms. This permits us to generate complex blend functions based on potentials. Using the weighted average of the potential functions for two objects permits us to define a continuous deformation from one to the other (Figure 4.20).

In Figure 4.20 the upper images show solids (spheres, cubes, and helices) averaged with each other in varying proportions. The lower images show a sphere averaged with pseudo-fractal solid texture in a similar way. The ability to see solid texture as a solid may not yet have many applications but illustrates our desire to harmonize intuitively different types of modelled entity.

4.6.2 WINSOM90

WINSOM90 is a refined system that consolidates the concepts of WINSOM and WINSOM II. It is a response to the increasing need for application development, and also addresses the trend away from links between particular hardware and software and towards portability across a variety of platforms.

In WINSOM90, the original WINSOM concepts are restructured and re-expressed as compatible independent functions accessible through an *A*pplication *P*rogramming *I*nterface (API). The API is generally *state-free*, a characteristic which minimizes the imposition of constraints on the application programming environment, and which renders it particularly well-suited for use in object-oriented designs.

In an extension of the WINSOM II philosophy, WINSOM90 seeks to harmonize not only texture and solid definitions, but also other modelling concepts, such as transforms, lights and cameras.

4.6.3 Animation

Despite having produced several successful animations, including the ones mentioned in this article as well as those described elsewhere, we are aware of shortcomings in our animation capability. WINSOM does not utilize any interframe coherence, since we currently know no way to exploit this reliably [45,46]. We are, however, exploring the possibilities for running WINSOM on parallel processors, and we are collaborating on an implementation that runs on an array of Transputers™ [47].

In any case, raw frame-generating times are not the only problem. The difficulty is that of planning and choreographing an animation effectively. Shortcomings in this respect often surface at frame-generation time, when sequences need to be repeated! We have made animation less painful by designing a computerized storyboard system compatible with WINSOM and, more importantly, with ESME.

Animation technology is evolving rapidly. Most WINSOM animations have been produced by independently rendering up to 25 separate images for each second of animation time. Video tapes have then been constructed using frame-buffers and high-resolution video-recording systems. More recently, 35mm film systems have been used to record animation sequences of cinematographic quality.

Alternatively, high-function graphics workstations with inbuilt surface rendering facilities can now be used to rapidly construct and display images of quality acceptable at least for preview and investigative purposes, and, in some cases, for final production purposes. When these are used in conjunction with video-disc technology, the process of creating animations can be greatly accelerated.

ACKNOWLEDGEMENTS

We are grateful to everyone who worked with us on the WINSOM projects, both those from non-IBM institutions and research Fellows at the IBM United Kingdom Scientific Centre (UKSC). In particular, we would like to mention collaboration with the following: on the static modelling of magnetic monopoles—A J G Hey, J H Merlin (Southampton University) and M T Vaughan (Northeastern University, Boston); on the dynamic modelling of magnetic monopoles—M F Atiyah, N J Hitchin (Oxford University), J H Merlin (Southampton University), and D L Pottinger (UKSC Fellow); on animation of molecular-beam epitaxy—D D Vvedensky and S Clarke (Imperial College, London); on the reconstruction of hippocampal neurons—H V Wheal (Southampton University) and H M Cole (UKSC Fellow); on the modelling of liquid crystals—G R Luckhurst and D J Tildesley (Southampton University); on geometric algebra systems—J H Davenport and A Bowyer (Bath University); on the parallel processing version of WINSOM—P M Dew and D Morris (Leeds University); on the modelling of proteoglycan assemblies—K Parker, P Winlove (Imperial Cancer Research Institute), M Bayliss and T Hardingham (Kennedy Institute); on all of the UKSC projects—students from various universities and polytechnics.

REFERENCES

1. P. Quarendon, *WINSOM User's Guide*, IBM UK Scientific Centre Report 123, 1984.

2. J.R. Woodwark and K.M. Quinlan, "Reducing the effect of complexity on volume model evaluation," *Computer-Aided Design* **14**(2), 89 (1982).
3. J.R. Woodwark and K.M. Quinlan, "The derivation of graphics from volume models by recursive subdivision of the object space," *Proceedings of Computer Graphics 80 Conference*, Brighton, Online Publications, Ltd., (1980), p. 335.
4. H. Samet, "The quadtree and related hierarchical data structures," *ACM Computing Surveys* **16**(2), 187 (1984).
5. A.R. Halbert and S.J.P. Todd, *Fast Redundant Primitive Removal in CSG Processing*, IBM UK Scientific Centre Report 175, in preparation.
6. J.R. Woodwark, "Generating wireframes from set-theoretic solid models," *Computer-Aided Design* **18**(6), 307 (1986).
7. A.R. Halbert and S.J.P. Todd, *FastDraw: A Fast Drawing System for Constructive Solid Geometry*, IBM UK Scientific Centre Report, 179, in preparation.
8. A.S. Glassner, "Space subdivision for fast ray tracing," *IEEE Computer Graphics and Applications* **4**(10), 15 (1984).
9. J.R. Woodwark and A. Bowyer, "Better and faster pictures from solid models," *Computer-Aided Engineering Journal* **3**(1), 17 (1986).
10. R.G. Oliver, *Hierarchical Data Structures for Solid Modelling*, Ph. D. thesis, University of Leeds, 1987.
11. S.J.P. Todd, "The Peterlee relational test vehicle—A system overview," *IBM Systems Journal* **15**(4), 285 (1976).
12. T.R. Heywood, B.N. Galton, J. Gillett, A.J. Morffew, P. Quarendon, S.J.P. Todd, and W.V. Wright, "The Winchester Graphics System: a technical overview," *Computer Graphics Forum* (Proceedings of Eurographics UK 84, April 1984) **3**, 61 (1984).
13. A.J. Morffew, S.J.P. Todd, and M.J. Snelgrove, "The use of a relational database for holding molecule data in a molecular graphics system," *Computers and Chemistry* **7**(1), 9 (1983).
14. J.M. Burridge, A.J. Morffew, and S.J.P. Todd, "Experiments in the use of PROLOG for protein querying," *Journal of Molecular Graphics* (Proceedings of the Molecular Graphics Society 4th International Meeting, April 1985) **3**(3), 109 (1985).
15. A.J. Morffew and S.J.P. Todd, "The use of PROLOG as a protein querying language," *Computers and Chemistry* **10**(1), 9 (1986).
16. E.E. Abola, F.C. Bernstein, S.H. Bryant, T.F. Koetzle, and J. Weng, "Protein data bank," in **Crystallographic Databases**, F.H. Allen, G. Bergerhoff, and R. Seivers, eds., (Data Commission of the International Union of Crystallography, Bonn, 1987).
17. S.J.P. Todd, *ESME, an Extensible Solid Model Editor*, IBM UK Scientific Centre Report 176, in preparation.

18. J.J. van Wijk, "SML: A solid modelling language," *Computer-Aided Design* **18**(8), 443 (1986).
19. J.R. Woodwark, "Eliminating redundant primitives from set-theoretic solid models by a consideration of constituents," *IEEE Computer Graphics and Applications* **8**(3), 38 (1988).
20. J.R. Rossignac and H.B. Voelcker, "Active zones in constructive solid geometry for redundancy and interference detection," *ACM Transactions on Graphics* **8**(1), 51 (1989).
21. A.R. Halbert, S.J.P. Todd, and J.R. Woodwark, "Generalizing active zones for set-theoretic models," *Computer Journal* **32**(1), 86 (1989).
22. A.F. Wallis and J.R. Woodwark, "Interrogating solid models," *Proceedings of the CAD 84 Conference*, Brighton, Butterworths (1984), p. 236.
23. A.F. Wallis and J.R. Woodwark, "Creating large solid models for NC toolpath verification," *Proceedings of the CAD 84 Conference*, Brighton, Butterworths (1984), p. 455.
24. A. Berk, L. Brownston, and A. Kaufman, "A new colour naming system for graphics languages," *IEEE Computer Graphics and Applications* **2**(5), 37 (1982).
25. S.J.P. Todd, *Winchester Colour and Texture Facilities: WINCAT*, IBM UK Scientific Centre Report 203, in preparation.
26. S.J.P. Todd, *SLED, A Solid Lighting Editor*, IBM UK Scientific Centre Report 177, in preparation.
27. K. Perlin, "An image synthesizer," *ACM Computer Graphics* (Proceedings of SIGGRAPH 85) **9**(3), 287 (1985).
28. R.W. Floyd and L. Steinberg, "An adaptive algorithm for spacial grey scale," *SID 75 Digest*, 36 (1975).
29. S.J.P. Todd, *Error Diffusion for Grey Scale and Colour Images: How to Make It Work*, IBM UK Scientific Centre Report 180, in preparation.
30. J.M. Burridge, A.J. Morffew, P. Quarendon, and S.J.P. Todd, "WGS: a flexible combination of interactive and solid molecular modelling," *UCLA-GENEX Symposium on Protein Structure, Folding, and Design*, (1985).
31. J.M. Burridge and S.J.P. Todd, "An interactive interface for protein secondary structure representation," *Journal of Molecular Graphics* **2**(53) (1984).
32. P. Quarendon, *A System for Displaying Three-Dimensional Fields*, IBM UK Scientific Centre Report 171, 1987.
33. P. Reilly and A.R. Halbert, *Using Computer Graphics to Analyse Archaeological Survey Data from the Isle of Man*, IBM UK Scientific Centre Report 153, 1987.
34. P. Reilly, "Data visualization in archaeology," *IBM Systems Journal* **28**(4), 569 (1989).

35. H.M. Cole, T.R. Harris, E.W. Stockley, and H.V. Wheal, "Interactive three-dimensional display of reconstructed neurons," *Journal of the Physiological Society* **400**, 5 (1988).
36. J. Simmons and A.B. Johnson, *Obtaining 3-Dimensional Models of Dendritic Structures*, IBM UK Scientific Centre Report 202, in preparation.
37. A.J.G. Hey, J.H. Merlin, M.W. Ricketts, M.T. Vaughn, and D.C. Williams, "Topological solutions in gauge theory and their computer graphic representation," *Science* **240**, 1163 (1988).
38. W. Latham and S.J.P. Todd, "Visualization of sculpture," *Proceedings of a BCS Displays Group Conference on Realism and Visualisation*, London (1987).
39. A. Ricci, "A constructive geometry for computer graphics," *Computer Journal* **16**, 157 (1973).
40. A.H. Barr, "Global and local deformations of solid primitives," *Computer Graphics* **18**(3), 21 (1983).
41. J.R. Woodwark, "Blends in geometric modelling," *Proceedings of 2nd IMA Conference on the Mathematics of Surfaces*, Oxford University Press (1986), p. 255.
42. D.-Y. Zhang and A. Bowyer, "CSG set-theoretic solid modelling and NC machining of blend surfaces," *Proceedings of the ACM 2nd Symposium on Computational Geometry*, Yorktown Heights, NY (1986), p. 236.
43. D.-Y. Zhang, *CSG Solid Modelling and Automatic NC Machining of Blend Surfaces*, Ph. D. thesis, University of Bath, 1986.
44. A. Bowyer, J. Davenport, P. Milne, J. Padget, and A.F. Wallis, "A geometric algebra system," in **Geometric Reasoning** (Proceedings of a Conference at the IBM UK Scientific Centre, 1986), Oxford University Press, 1989.
45. A.S. Glassner, "Spacetime ray tracing for animation," *IEEE Computer Graphics and Applications* **8**(3), 60 (1988).
46. J.R. Woodwark, "Spacetime ray tracing," *IEEE Computer Graphics and Applications* **8**(5), 8 (1988).
47. N.S. Holliman, D.T. Morris, P.M. Dew, and A. de Pennington, "An evaluation of the processor farm model for visualising constructive solid geometry," *Proceedings of Computer Vision and Display*, Leeds University Conference, to appear.

5

Animations and Graphics in Molecular Dynamics Simulations

Per Stoltze

Some aspects of animation of Molecular Dynamics simulations are presented and the design and use of animation programs for this purpose is discussed. The discussion is limited to some aspects that have actually been implemented and found useful. Animations in connection with simulations are mainly used for analytical purposes and for presentation of the results. Interactive and flexible production of the animation is important in particular for analytical applications.

5.1 INTRODUCTION

In recent years the cost of large-scale computation has dropped dramatically and the systems for computer-generated graphics have become much more powerful and accessible. Parallel with these developments in hardware and software, excellent textbooks on computers [1], computer science [2–6], numerical analysis [7,8] and graphics [9] have become available.

Even if there is no doubt that the use of animations in scientific computations is in its very beginning, the developments in computing have already changed the way simulations are carried out. Countless new applications of graphics and animation in computer simulations can be imagined. The following account will be limited to some aspects of animations that we have actually implemented and found useful in connection with the analysis and presentation of Molecular Dynamics simulations of some problems in surface science.

5.1.1 The Purpose of Simulations

Simulations have an intermediate role between total-energy calculations and experiment. While total-energy calculations can give detailed prescriptions for the calculation of interactions between atoms, there are aspects of atomic interactions which are difficult to answer by present day total-energy calculations: high temperatures, disordered systems and many-atom dynamics are some examples. Analogously, there are limitations to experiments. One is limited to systems that one can actually prepare with respect to structure, defects and purity.

The starting point for the simulations are the prescription for the calculation of interaction energies obtained from theory. From experiments we need reliable data that can be used to investigate if the simulations are reproducing the essential aspects of the physical problem under investigation. Once the simulations have been improved to the limit of agreement with experiments, the simulations may contribute to theory, *e.g.*, by investigations of the macroscopic consequences of the microscopic interactions. Furthermore, the simulations can contribute to experimental studies by the calculation of properties that are well-defined but cannot be measured. The structure of atomic layers near the surface of a metal crystal is an example.

Aside from contributions to theory and experiment, the output from simulations can be used as the starting point for animations. These animations are important for the formation of an intuition about the phenomenon as well as for the presentation of the results.

As for pure theory, computers are the most important equipment used in simulations. However, simulations resemble experiments in the respect that the results are subject to both statistical and systematic errors. Further, the high resolution in simulations is not always an advantage. Whenever we perform a simulation, we must be aware that the full complexity of the physical system will be reflected in the results. A few artifacts, *e.g.*, from finite-size effects or from the approximate nature of the potential may also be present.

5.1.2 The Purpose of Graphics

Graphics and animations have several different applications in simulations. One important application is to examine the output from simulations for convergence problems, to screen output for interesting observations and to build an intuition for the phenomenon under study. For the inspection of simulation output by animation it is important that the graphical representation is fast and flexible. We want to be able to ask questions and get the answers while sitting at the workstation. *Black and white* pictures are often sufficient, but the possibility of labelling the atoms by numbers (position in input file, layer, coordination, energy, *etc.*) is necessary.

Another important application of animation is to present the results of simulations. The objective of the animation for presentation purposes is quite different from the use of animation for analytical purposes. The animation should primarily be instructive and clear. This implies that coding the atoms by colours can be quite effective, while labelling by numbers or letters will not work. For presentations we will probably want to record the animations on an ordinary video cassette with short texts between different sequences. The turnaround for the production of the animation may be longer than that for the animation for analytical purposes.

5.1.3 Some Words of Caution

The human mind has a remarkable ability to capture correlations and changes in visual patterns [10,11]. It is our ability to perceive dynamic patterns that make animations so useful. However, we should be aware that the perception of correlations is *not at all* quantitative. The perception of movements we get from watching white noise on a television screen is a reminder that we think we see correlations in dynamic patterns even when no correlation exist [10]. For the application of animations in the study of correlations we should be aware of the qualitative nature of our perception of correlations. Watching animations cannot replace the quantitative study of the correlations.

For projections there is an ambiguity in the interpretation of the spatial arrangement. This phenomenon is illustrated in Figure 5.1, where one cannot tell if the gray side of the box is at the top or the rear. If the system is slowly rotating, the axis of rotation or the direction of the rotation may be ambiguous. While this can be irritating to watch, it is less serious than the ambiguity in the interpretation of projection of more irregular objects, simply because in the latter case the ambiguity is not perceived. If the system is slowly rotated, the perception of the spatial arrangement is much better. The use of a perspective rather than a projection reduces the ambiguity significantly. The ambiguity can be completely removed if stereo images are used [12].

5.2 PROGRAMS

In this section we will briefly discuss some aspects of simulations that are relevant for animations. Typically we will perform simulations by generating a sequence of configurations for the system under study and then calculate kinetic or thermodynamic data by analysis of these configurations. A number of programs cooperate in the generation and analysis

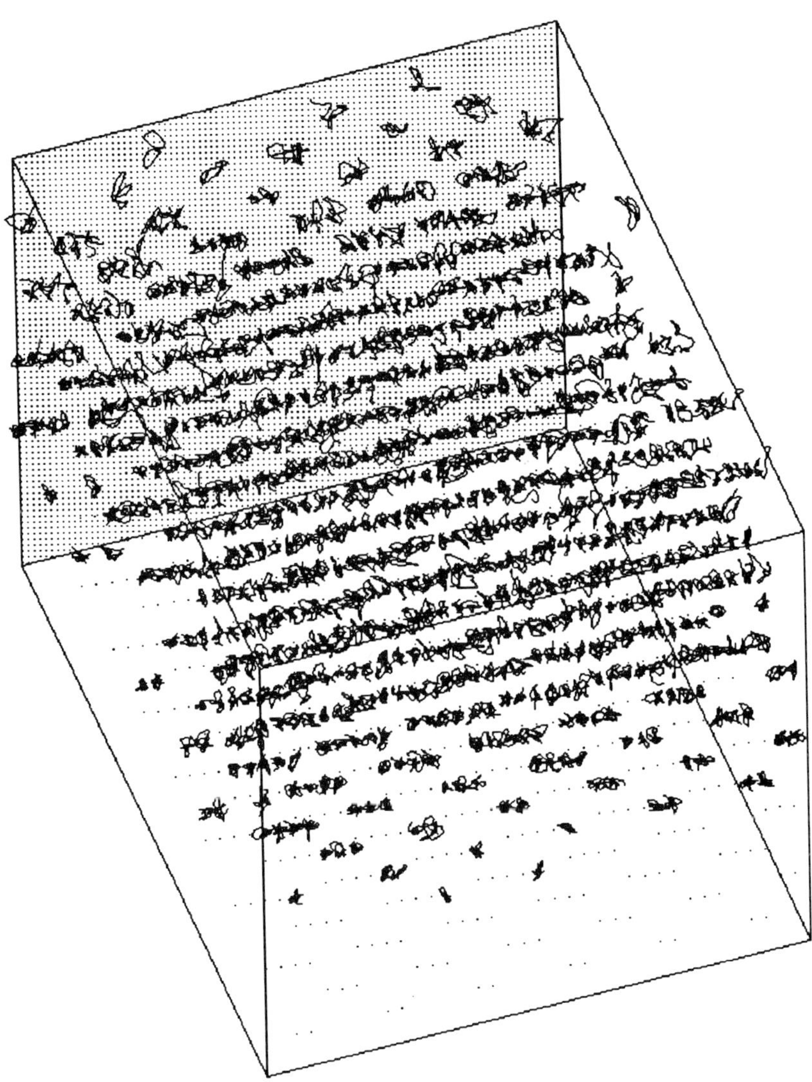

Figure 5.1 The trajectories of atoms displayed as a trace. The box illustrates the period of the system and one side of the box has been coloured gray. It is not possible from the picture to say if the coloured side is at the rear or at the top of the box. Actually, this picture is a frame from an animation showing a slow rotation of the system. In the animation it is obvious that the gray side is at the rear.

of large amounts of data. In the following sections we will give a brief exposition of how we have chosen to implement such a system.

5.2.1 Setup

The starting configuration[†] for the simulation is either the final configuration of a previous simulation or a configuration generated by a setup-program. The system may be a three-dimensional lattice, a slab or a cluster of a specific structure. The surfaces have specific orientations and may have defects.

5.2.2 The Interaction Potential

The choice of interaction potential is very important for the realism of the result of the simulations, but it has little connection to the problems of animation with one important exception. The *garbage in, garbage out* principle also applies to animation programs, implying that if the simulations are not realistic, they cannot be saved by a fancy presentation.

5.2.3 The Principle of Molecular Dynamics

The principle in Molecular Dynamics [13,14] is that the equations of motion for the system are integrated. The averages are calculated as averages over time. The advantage of Molecular Dynamics is that time is well-defined and dynamic quantities, such as the diffusivity, may be calculated.

5.2.4 The Principle of the Monte Carlo Algorithm

The principle in the Monte Carlo [13–16] is that a stochastic algorithm is used to generate configurations according to a prescribed distribution. The advantage of the Monte Carlo approach is that we only need to calculate the energy, not the force. Usually the energy calculation is somewhat easier than the force calculation. For some situations the force does not exist; for instance, it is not easy to formulate ordinary thermal expansion in terms of a force acting on the lattice constant.

However, time is not well defined in a Monte Carlo simulation and the animation of the output will show the development of the system in "Monte

---

[†] We will use the word *configuration* to indicate the physical state of the system. In simulations and, in particular, in animations there are some additional parameters, *e.g.*, the viewing angles in an animation. We will use the word *frame* to refer to a configuration with definite values for these computational parameters.

Carlo time." This may or may not have a simple relation to the development of the system in physical time.

5.2.5 Data File Format

One does not want to store all the configurations generated during the simulation. Usually one wants to store 2-4 frames per lattice vibration in Molecular Dynamics. In Monte Carlo one will usually store frames when the number of accepted moves equals one per degree of freedom in the system. If frames are stored more frequently, they may show a rather high correlation.

The configurations are stored in a data file of a chosen format. The same format is used for the output from Monte Carlo and Molecular Dynamics. The file contains a header and a block of data for each configuration. The header contains a headline, the number of atoms, the translational symmetry, the time between configurations, the number of configurations in the file, *etc.* For each configuration we store the basis vectors and for each atom its name, position, momentum, force, energy, *etc.* The storage of information should depend on the balance between the work involved in the storage (CPU-time for output, increased disk space, additional CPU-time spent each time the file is read) compared to the work involved in the recalculation of some data.

In addition to the name, position, *etc.*, for each atom we have left space for one integer for each atom, corresponding to its class. The output can then be manipulated by a number of programs and analyzed by other programs. The class of each atom is used in the communication between the programs. The programs that manipulate the file, do so *exclusively* by modifying the class of the atoms. We can thus undo any manipulation carried out on the file by resetting the class of each atom to its default value. The class can be set equal to the layer number for the atom, its coordination number or it may be interpreted as the colour of the atom by an animation program. The class "0" refers to atoms that should be ignored or treated as invisible.

5.3 PROGRAM DESIGN

Even if programming and, in particular, computer animation may be great fun, the objective is to have the shortest possible time from when the physical problem is formulated to the time when the final answer has been found. Considerable amounts of time may be spent on the design, coding, debugging and verification of programs. The number 50% has been quoted for the percentage of the total time spend on debugging and verification of computer programs [17], and 95% as the percentage of large programs that are never finished or never used [18]. Add to this the requirement that we want to do physics and that the programming is not in itself the goal, and we are facing a huge problem.

5.3.1 Development Tools

A collection of programs for simulations, analysis and animation of the results is quite large and the development of software should be handled accordingly. Programming tools such as the UNIX tools *sccs* for control of program revisions, *make* for control of compilations, *lint* for program verification and *dbx* for source level debugging are all indispensable for projects of this size.

We have found that the *awk* language [19,20] is very powerful when used interactively for extraction of data, compilation of databases and for exploratory data analysis. Furthermore, we need good tools for *xy*-plots and for the editing and display of raster images. For these purposes *grtool* [21], *touchup* [22] and *movietool* [23] have passed the test of time.

5.3.2 Language

The speed of program development seems to be higher if high-level languages are used [17]. For this reason we have chosen to implement simulation and graphics in FORTRAN-77 [24] for computer-intensive subtasks, C [2,25] for other tasks and PostScript [26,27] for graphics. All are highly portable. PostScript is a *de facto* standard high-level graphics language, and offers a large number of useful features, including a well-chosen set of commands, possibilities for recursion and iteration and good facilities for the integration of graphics, text and raster images.

5.3.3 Structure

The distribution of the tasks carried out in the course of the analysis into a number of programs offers several advantages. A distribution almost to the limit of one step in the analysis performed by one program, one task performed by one subroutine, allows the use of different design objectives and widely different algorithms in different analyses. Furthermore, a program performing one step in the analysis using one algorithm is simpler to design and faster to write and debug. Finally, one can use different level of sophistication during the lifetime of a program. A program used only a few times may be put together starting from existing programs. Advanced algorithms can then be put in if the need arises and the verification can be done incrementally.

5.3.4 Hardware

One passing remark about hardware. The objective in the choice of hardware is not the fastest computational speed, but the fastest turnaround from submission of the simulation to the completion of the analysis. Rather than comparing the *peak performance* or the *sustained performance* of the hardware it may be more relevant to compare the sustained performance divided by the number of users. The rapid development in the fields of fast workstations and massively parallel supercomputers is a further argument for writing highly-structured, portable programs.

5.3.5 Examples of Analysis Programs

Programs used in simulations may be separated into *data manipulation programs* used to prepare the output for analysis by other programs, *analysis programs* performing various analyses and *graphical programs* for graphics and animation.

Some examples of *manipulation programs* are programs that sets the class of an atom equal to the layer number for the atom, sets the class of each atom equal to the coordination number for the atom, or makes immobile atoms invisible.

The *analytical programs* calculate various thermodynamic, structural or kinetic data. Using the manipulation programs, each of the analytical programs can be used to calculate data for the system as a whole, for each layer, for each coordination number, *etc.* Among the analytical programs are programs that calculate histograms of atomic positions, of energies and of structure factors, calculate the development in energies, both as averages for each configuration and as a running average, calculate the diffusivity, calculate power spectra, Fourier transforms, *etc.* Other analytical programs may test for specific problems in the simulations.

5.4 ANIMATION

Animation can be used as an analytical tool if it is used with the data manipulation program to label the atoms. As an example, one may label the atoms by numbers to study the mechanism of diffusion. Using other programs to make immobile atoms invisible or to label the atoms by energy can help the analysis.

Animations are not limited simply to sequences of the motion of atoms. As another example, the development of a correlation function may be animated. In addition, animations are not limited to the display of developments in time. For example, one may plot the density profile for the system and display results as a temperature scan. The limit between graphics and animation is thus dissolving as we may animate data other than atomic positions or present the motion of atoms as a still-life, as discussed in Section 5.4.6.

If we want to have a good view of the system it is obviously much easier to rotate the system during the generation of the animation rather than starting the system in a particular orientation or even making it rotate during the simulation. The animation program must thus be able to rotate the system for display purposes, either by showing the entire animation at a particular viewing angle or by scanning through an interval of viewing angles.

The periodic boundary conditions of the system may be used to *replicate* the system to give a better presentation. Also, if the system is slowly rotating, we may *repeat* each configuration to make the rotation more smooth. For example, we might use each configuration twice to present 30 configurations as 60 images using a 6° rotation between each image.

The atomic positions may be presented as a *perspective* or as a *projection*. While the spatial arrangement is more obvious in a perspective, the comparison of projections is easier. As data analysis is as important as the presentation, we prefer to make the animations as projections.

Before discussing the design of animation programs in the next section we will discuss some problems in animations.

5.4.1 Correlation between Frames

In simulations we wish to retain a high correlation between atomic positions in any two consecutive frames. This implies that we should have a quite high correlation (50–80%) between consecutive pairs of stored configurations. This is somewhat higher than optimum for the calculation of correlations in time and much higher than the optimum (0%) for the determination of averages.

If the correlation is too low it becomes very difficult to follow individual atoms, with the result that the trajectory of an individual atom tends to be perceived incorrectly. This happens as the mind confuses the identity of atoms by assuming that the displacements of atoms are shorter than they are in reality. This problem may be solved in a number of ways. For animations made for analytic purposes, it may be best to label the atoms by numbers and watch the animation at a very low speed, *e.g.*, 0.2 frames per second. For display purposes, it will be better to apply colour coding to make the atoms distinguishable. At the beginning of a project it may be a good idea to calculate the autocorrelation from a trial run and use this information to determine the optimum sampling rate to be used in the simulations.

5.4.2 Boundary Conditions

For systems with periodic boundary conditions the boundary effects may be quite dominating in the animation if just a single instance of the system is shown. It may be better to use the translational symmetry of the system to replicate it a few times. The periodicity of the resulting structure is often not too obvious.

If atoms move over large distances in systems with periodic boundary conditions, we have to decide on a graphical treatment of this motion. If we choose to show the system or just a few replications of it, with no special treatment of the diffusing atom, this atom may appear to leave the system. This gives a quite wrong impression of the situation. Another possibility is to use the periodicity to shift the atom back in from the opposite side of the system at the moment it would otherwise appear to leave the system. However, this is not a very good solution as the mind immediately detects this as a flicker at the boundary of the system and the process will attract too much attention.

5.4.3 Economy of Plotting Commands

A picture can be very large if it is written as plotter commands. A command such as *p 243 423* is 9 bytes and may only flip one pixel in the output. Even if we use a very compact notation the commands to plot nearest neighbour bonds in a 2000 atom configuration may exceed 512 kB, and the commands to plot a trace of 2000 atoms through 300 frames may exceed 4.8 MB. A conclusion is that the programs used for graphics and animation should be able to handle an unlimited number of commands, or at least between 10 and 100 MB of commands depending of the notation used.

5.4.4 Size of Images

For a presentation we may use quite large raster images, *e.g.*, 1152× 900 pixels. Storing such pictures on disk [12] requires approximately 13 kB for a byte encoded monochrome image (Figure 5.2), approximately 35 kB for the same image stored as a 8-bit colour image. Traces take up more storage, approximately 28 kB for Figure 5.3 stored as a 1152×900 pixel, monochrome image. For images to be used in use smaller images, *e.g.*, 300×200 pixels, the images are more complex and 6 kB will be typical for a byte-encoded monochrome image. Storage without some kind of compression is almost impossible, the 35 kB 8-bit colour image mentioned above will consume 1.04 MB, if stored without compression.

Raster images is not the only form for storage of pictures, in Section 5.5.1 we will discuss the storage requirements for pictures stored as PostScript.

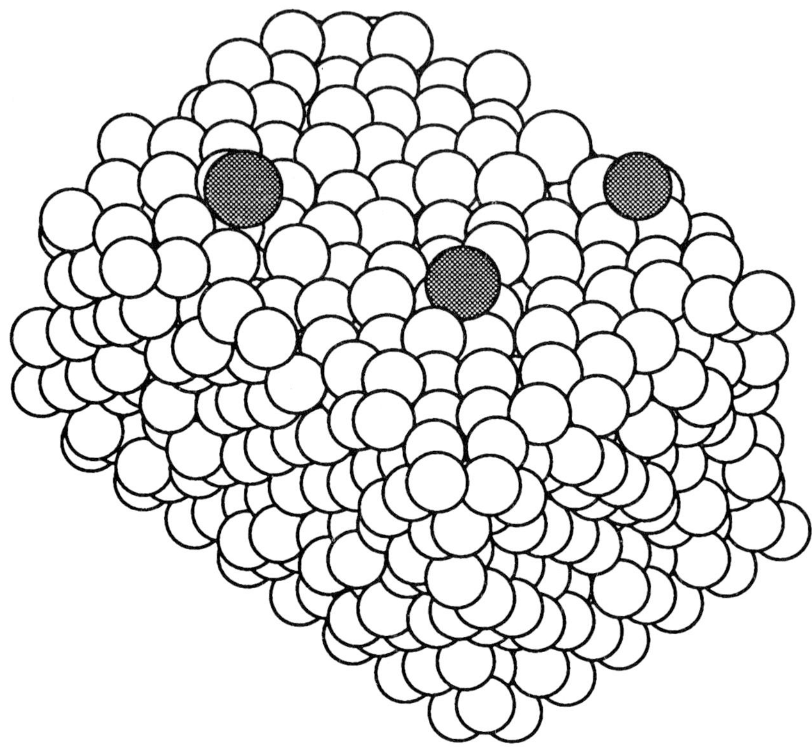

Figure 5.2 Atoms displayed as opaque circles. Contrary to more realistic displays (Figure 5.5) plain circles can be used as a basis if the atoms are to be labelled. In the figure, the adatoms floating on the top of the surface have been coloured gray.

5.4.5 Playback Speed

The choice of playback speed is a compromise. One may choose to play 16 frames per second. If the correlation between pairs of consecutive frames is not too low, this will give an excellent presentation, but it will also require a quite large storage for a sequence of fixed length, *e.g.*, 30 seconds, and loading the images into the RAM may be impossible. We could circumvent this difficulty by using *single-frame recording.* However, this technique is by nature quite slow and the high cost of the equipment tends to reduce the turnaround for the animation even further. Single-frame recording may

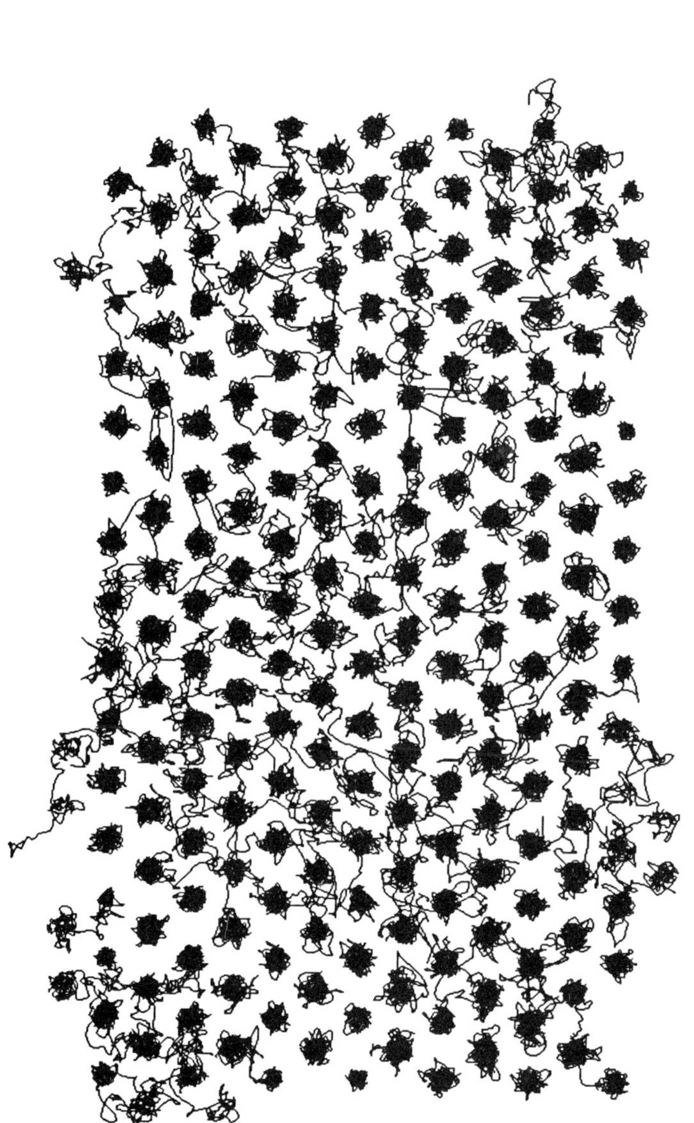

Figure 5.3 The atomic positions in a single layer of atoms displayed as a trace. In this presentation spatial details are retained, but correlations in time are entirely lost.

be used for animations for presentation purposes, but is it not at present an option in analytical work.

Deleting every other frame and playing the animation at 8 frames per second will of course reduce the required storage by a factor of 2. However, if the original animation had an optimum correlation between frames, the speed-up will cause some flicker. Playback at a lower speed than approximately 8 images per second is not acceptable. There are several technical solutions to these problems [12]. As long sequences are not too interesting and as speeding up the display will cause both flicker and storage problems, a solution that should not be forgotten is to represent very long sequences as a series of still pictures showed at 5–7 second intervals.

5.4.6 Presentation on Paper

While playback on a workstation is sufficient for everyday use and recording on a VCR is preferable for presentations, one may have to make a presentation on paper. One may choose to present the results as traces, as shown in Figure 5.3, or as comic strips, as shown in Figure 5.4. The advantage of comic strips is that temporal resolution is retained and the frames can be superimposed for comparison. This can be done very effectively either by using a raster editor, such as *touchup*, or by printing the comic strips on transparent foil. In the comic strip the geometric details are rather hard to see and all interior details will be hidden. The trace represent geometric details much better, but the temporal resolution is entirely lost. Also some aspects of the relative position of the atoms in space will be lost (Figure 5.1).

5.5 DESIGN OF ANIMATION PROCESS

The number of features that one might implement in programs for animation of simulations is unlimited. In the following we will limit the discussion to the features that we have actually implemented and found useful.

5.5.1 Display of Atoms as Spheres

A program for the presentations of atoms as spheres (Figure 5.5) can of course rotate and replicate the system, but contrary to the traces presented in the following section, the dynamics is limited to a scan through the configurations generated in the simulation. This display is mainly used for presentation purposes. However, for analytical purposes, this display of the atoms can be enhanced by a number of features to be discussed below.

The animation program reads the output from the simulation and a file containing the viewing angles, the size of the image, the details of

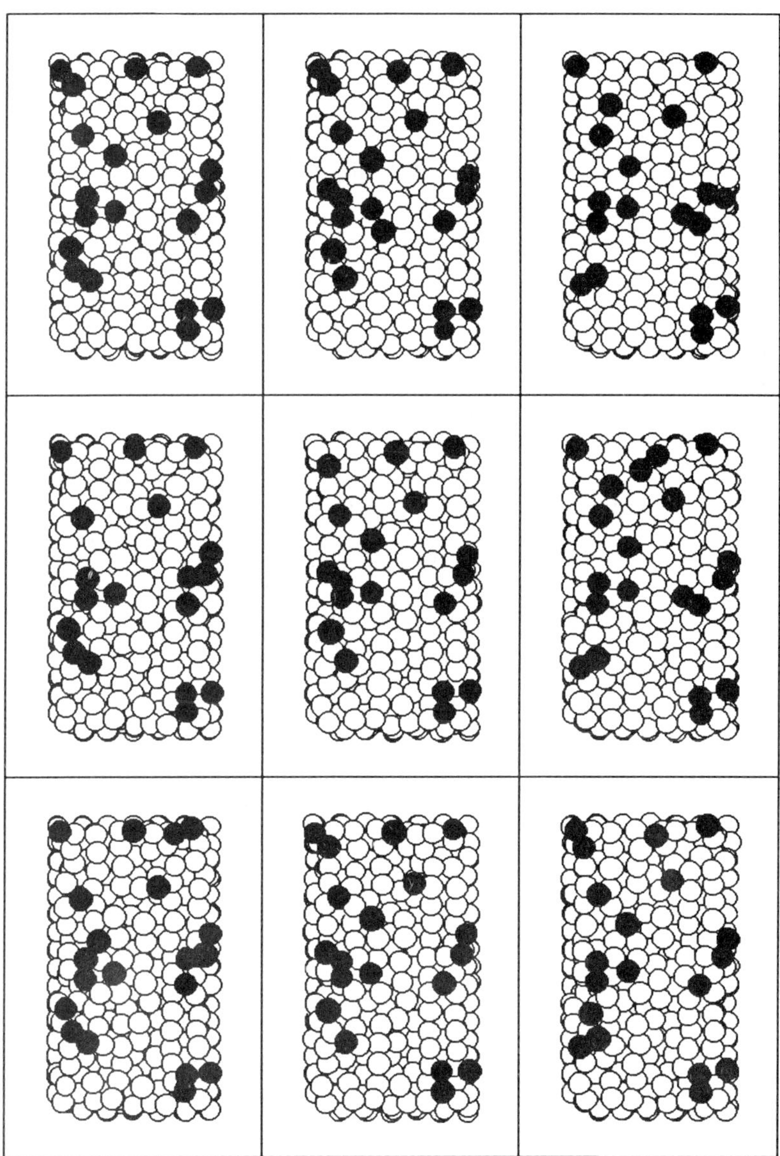

Figure 5.4 A series of frames from an animation assembled as a comic strip. In this display, the correlations in time are retained, but the amount of spatial details is much less than for the trace in Figure 5.3.

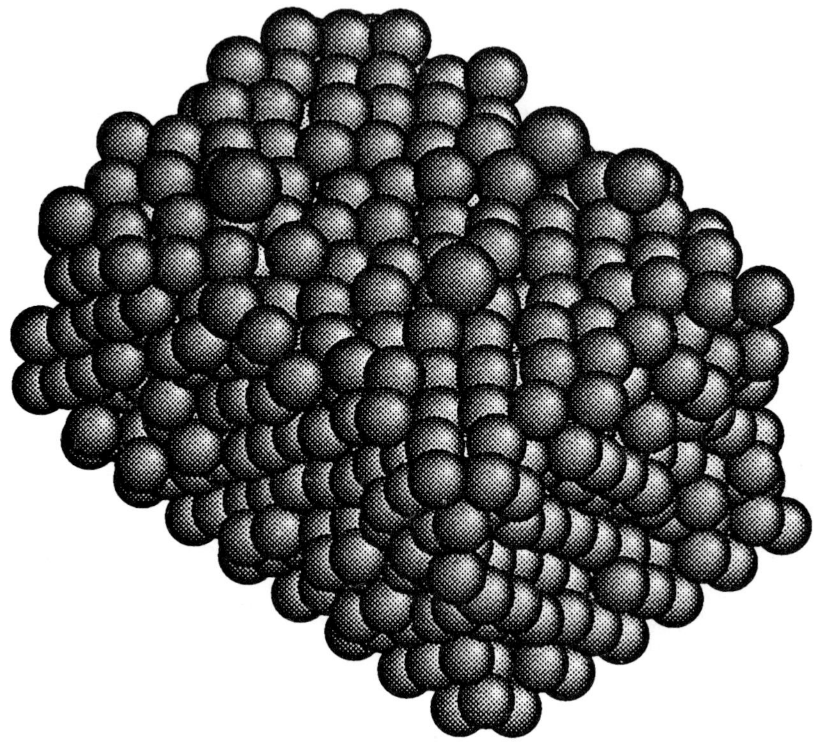

Figure 5.5 Atoms presented as shaded circles. This display is suitable for presentations, but it is not useful if the atoms are to be labelled by text or numbers.

the replication, *etc.* The immediate output from the simulations consists of PostScript. A prologue contains the macro definitions and each image is represented as one file containing coordinates and calls to the macros defined in the prologue. The details of the images may be adjusted by editing the prologue. Among the details that can be adjusted are the size and position of text, the colour of atoms, the line widths, the position on the paper and the colour of the background. Typical sizes of the output are 3.4 kB for the prologue and approximately 30 B for each atom after replication, *i.e.*, 30 kB for one picture of 1000 atoms.

The animation in the form of PostScript commands can be printed on paper or displayed on the screen of a workstation running a PostScript interpreter such as NeWS [28]. Displaying the PostScript commands as an animation is too slow, but PostScript can easily be converted by NeWS to raster images for playback by *movietool*.

Atoms may be displayed as circles (Figure 5.2) or spheres (Figure 5.5). The circles are drawn as an opaque disc with a black edge. *Opaque paint* is thus used as the technique for removing hidden details. The shaded spheres give a more natural appearance of the system, but the shaded spheres cannot easily be combined with other features such as labelling of the atoms by text.

A bond may be plotted between neighbour atoms if it is visible. The plotting of the bond greatly aids the perception of the structure. The display of the atoms as spheres of course obscures the interior of the system. However, the bonds offers the possibility of reducing the radius of the atoms to near zero to display the structure. Actually the radius should only be reduced to a few pixels to prevent isolated atoms from becoming invisible. The resulting figure clearly displays the lattice structure and in animations the lattice vibrations will be apparent.

For systems without a lattice structure, a display in the form of neighbour bonds is not instructive. Instead the structure may be displayed by reducing the radius of each atom to two-thirds of its true value, Figure 5.7. This makes it possible to form an impression of the structure, in particular if the system is slowly rotated.

Atoms may be labelled by text or by numbers (Figure 5.8). If the labels are written twice in two different colours offset by a few pixels, the text will be readable independent of the background for the text. Some examples of labels are the name of the atom, the layer number, the coordination number or some other datum for the atom, such as its energy or diffusivity.

5.5.2 Display of Atomic Trajectories

The trajectory of atoms may be displayed as a trace (Figure 5.1). The main application of this display is to show the development in time for analytical purposes, although the trace may be printed for presentations on paper.

As the amount of plotter commands for a trace is very large, we have chosen to implement the program in C using the Pixrect library [29]. The program can of course replicate and rotate the system. Displaying the period of the system as a box may aid the perception of the structure.

Contrary to the display of atoms as spheres we may show much more complex animations than just scanning through the frames. If $\mathbf{r}_p$ refers to the position of an atom in configuration number p in a simulation where N configurations were generated, one possibility is to display the entire trajectory by connecting $\mathbf{r}_p$ for $1 \leq p \leq N$ and scan through the viewing angles. Other possibilities are to show the development in time by connecting $\mathbf{r}_p$ for $1 \leq p \leq n$, while scanning through n or to show the displacement in a

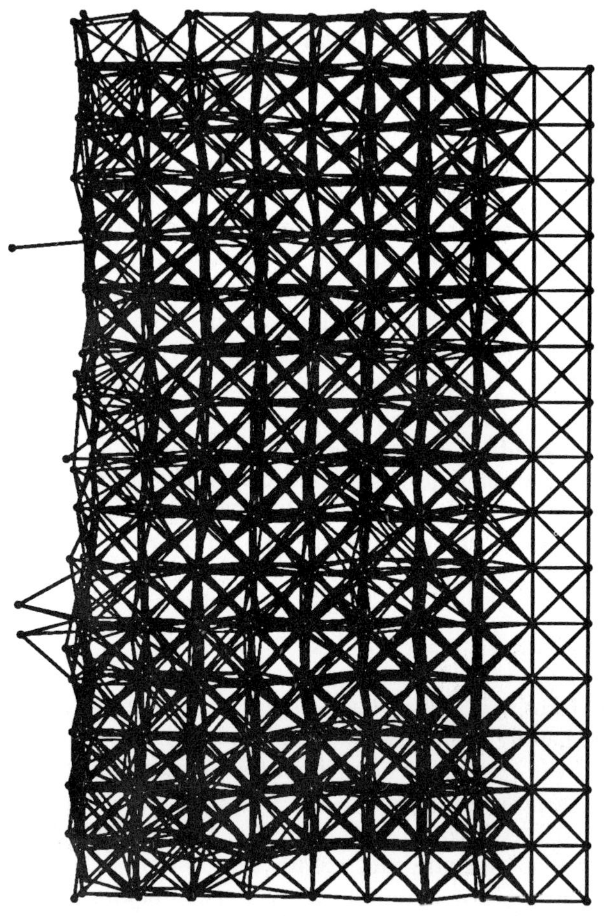

Figure 5.6 The radius of each atom reduced to near zero. The structure is then represented by a lattice of bonds.

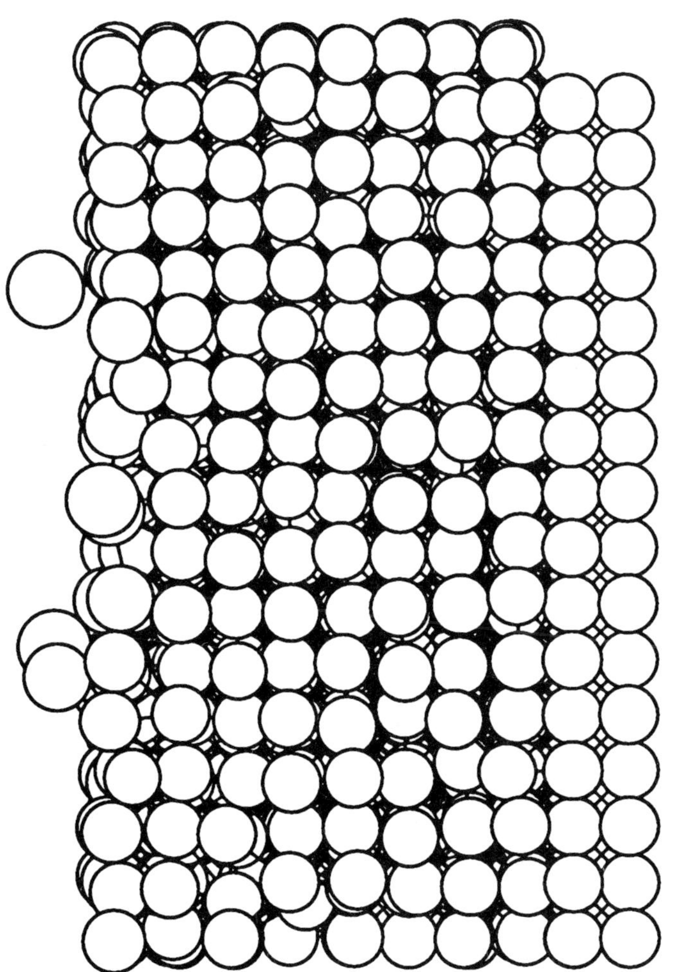

Figure 5.7 The radius of each atom reduced to two-thirds of the true value. This makes the structure of the system visible, in particular if the system is slowly rotated.

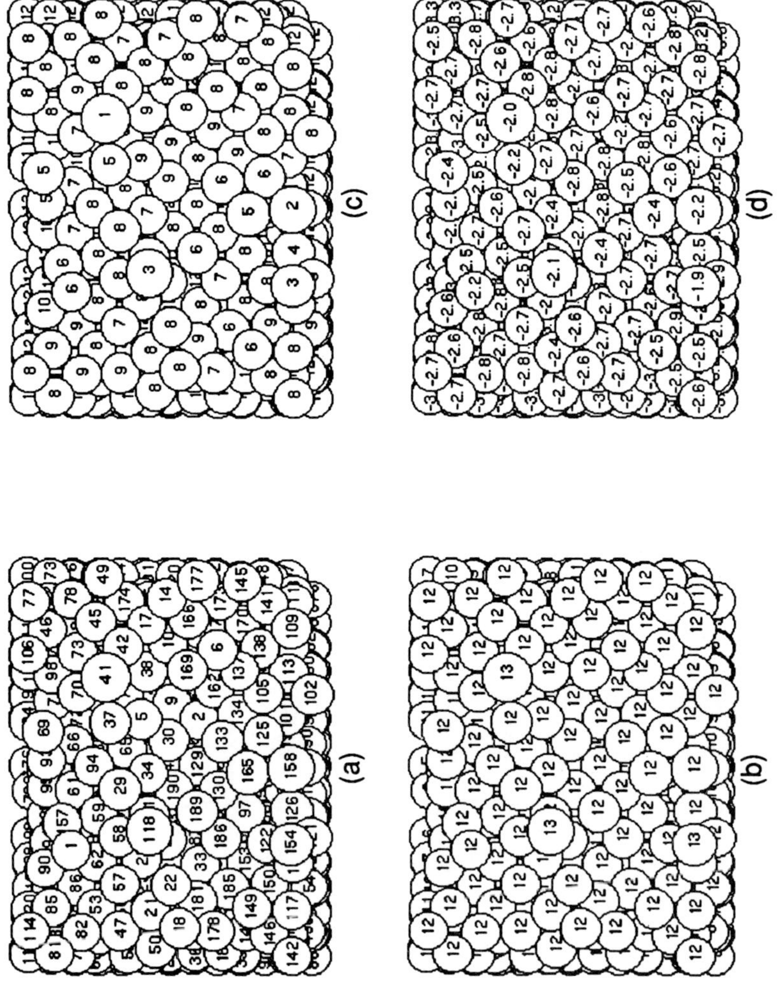

Figure 5.8 Fragments of the same configuration of atoms as in Figure 5.7, viewed from above, labelled by name (a), by layer (b), by coordinations (c) and by energy (d).

fixed time interval, m, by connecting $\mathbf{r}_p$ for $n \leq p \leq n+m$, while scanning through n. Other, more complex, displays related to various correlation functions may also be generated.

5.6 CONCLUSIONS

With the rapidly increasing possibilities for computer generated graphics, animations are becoming an important part of scientific computing. Additionally, the boundary between graphics and animations is becoming diffuse. Animations can be rendered as traces or as comic strips while other graphical information than atomic positions can be animated to show, for example, the development of correlation functions or a temperature scan. We have presented a brief account of some aspects of animations in connection with Molecular Dynamics simulations.

To complete the program development and verification as quickly as possible, we have chosen to use software development tools intensively. Analyses are implemented as a number of closely cooperating programs, with each program performing one operation, analysis or animation. Presentations in the form of animations are not quantitative and the perception of the motion has to be given some attention.

The main application of animations is to display information for analytical purposes or to present the results in an instructive way. These applications emphasize the easy and interactive preparation of the animations. Animations are made by displaying atoms as spheres or as traces of the trajectories. In the display of atoms, the possibility of displaying the structure by reducing the radius and the possibility of labelling the atoms are important. For traces, the important features are the slow rotation of the system and the possibility of showing quite complex information for analytical purposes.

ACKNOWLEDGEMENTS

Financial support from the Danish Research Councils through the *Center for Surface Reactivity* is gratefully acknowledged.

REFERENCES

1. K. Hwang and F.A. Briggs, **Computer Architecture and Parallel Processing** (New York, McGraw-Hill,1984).
2. B.W. Kernighan and M. Ritchie, **The C Programming Language** 2nd edition (Englewood Cliffs, NJ, Prentice-Hall, 1988).

3. H.M. Deitel, **An Introduction to Operating Systems** (Reading, MA, Addison-Wesley, 1984).
4. S.R. Bourne, **The UNIX-System** (Reading, MA, Addison-Wesley, 1983).
5. R. Sethi, **Programming Languages** (Reading, MA, Addison-Wesley, 1989).
6. R. Sedgewick, **Algorithms in C** (Reading, MA, Addison-Wesley, 1990).
7. W.H. Press, B.P. Flannery, S.A. Teukolsky and W.T. Vetterling, **Numerical Recipes in C** (Cambridge, Cambridge University Press, 1988).
8. W.H. Press, B.P. Flannery, S.A. Teukolsky and W.T. Vetterling, **Numerical Recipes: The Art of Scientific Computing** (Fortran and Pascal) (Cambridge, Cambridge University Press, 1986).
9. J.D. Foley and A. van Dam, **Fundamentals of Interactive Computer Graphics** (Reading, MA, Addison-Wesley, 1984).
10. V.S. Ramachandran and S.M. Anstis, "The perception of apparent motion," *Scientific American* **254**(6), 80 (1986).
11. V.S. Ramachandran, "Perceiving shape from shading," *Scientific American* **259**(8), 58 (1988).
12. O.H. Nielsen, "Animation in surface science," this volume.
13. M.P. Allen and D.J. Tildesley, **Computer Simulation of Liquids** (Oxford, Clarendon, 1987).
14. R.W. Hockney and J.W. Eastwood, **Computer Simulation Using Particles** (Bristol, Institute of Physics, 1989).
15. K. Binder (editor), **Monte Carlo Methods in Statistical Physics.** Topics in Current Physics **7** (Berlin, Springer, 1986)
16. K. Binder (editor), **Application of the Monte Carlo Method in Statistical Physics.** Topics in Current Physics **36** (Berlin, Springer, 1987).
17. F.P. Brooks, Jr., **The Mythical Man-Month** (Reading, MA, Addison-Wesley, 1982).
18. B.J. Cox, **Object-Oriented Programming** (Reading, MA, Addison-Wesley, 1986).
19. A. Aho, B.W. Kernighan and P.J. Weinberger, **The awk Programming Language** (Reading, MA, Addison-Wesley, 1988).
20. We have mainly used Sun Microsystems implementation of *awk*. Free Software Foundations *gawk* is a more complete implementation of the modern *awk* language [19]. *gawk* is available by anonymous FTP from `labrea.stanford.edu` [36.8.0.47].
21. *grtool* is a program for graphs and exploratory data analysis written by Paul J. Turner (`pturner@cse.ogi.edu`). *grtool* is available by anonymous FTP from `ess3.ese.ogi.edu` [129.95.20.62].

22. *touchup* is a raster image editor written by Ray Kreisel (rayk@sbcs.-sunysb.edu). *touchup* is available by anonymous FTP from `titan.rice.edu` [128.42.1.30].
23. *Movietool* written by Ole Holm Nielsen (`ohnielse@ltf.dth.dk`) animates a sequence of raster images by displaying them at a constant rate at a workstation. *Movietool* is available by anonymous FTP from `ltf.dth.dk` [129.142.66.16].
24. J.W. Crawley, and C.E. Miller, **A Structured Approach to Fortran** 2nd edition (Englewood Cliffs, NJ, Prentice-Hall, 1987).
25. A. Koenig, **C Traps and Pitfalls** (Reading, MA, Addison-Wesley, 1988).
26. Adobe Systems, Inc., **PostScript Language Manual** (Reading, MA, Addison-Wesley, 1985)
27. Adobe Systems, Inc., **PostScript Language Tutorial and Cookbook** (Reading, MA, Addison-Wesley, 1985).
28. Sun Microsystems NeWS is a network extensible window system based on a superset of the PostScript language. Alternatively one might use Free Software Foundations *ghostscript* interpreter available by anonymous FTP from `labrea.stanford.edu` [36.8.0.47] or Sun Microsystems OpenWindows.
29. Sun Microsystems *Pixrect graphics library* is a set of library routines for the manipulations of raster images by C programs.

6

Animation of Large-Scale Simulations

Mark R Wilby and Shaun Clarke

Interest in kinetic processes occurring over length-scales and time-scales that are not directly accessible to measurement has led to the development of computer simulations of these phenomena. This chapter describes the application of animation to simulations of a growth process known as molecular-beam epitaxy (MBE). The animations generated are transferred to domestic video, which allows an easy and flexible examination of fluctuations in large-scale simulations, where the enormity of the data produced usually forces the simulations to be monitored only in terms of averaged quantities.

6.1 INTRODUCTION

Computational power has brought about a revolution to the way we think about the world. With the improvement in the performance of computers, both in terms of speed and memory, we have been led into areas of research that, even a few years ago, would either not have been considered or would have been viewed in a completely different light. Computers have come to be looked upon as a tool, a playground, and a laboratory, becoming in effect a third avenue of scientific investigation, working alongside theory and experiment.

It could, however, be argued that in some sense computers havehindered progress. Because of the dramatic increase in computer performance in the past few decades, it is often unnecessary to tackle a problem from a variety of theoretical standpoints, but simply to seek the same type of solution, only using a bigger computer. As a particular numerical technique reaches its limits with current technology, new and faster technology provides renewed impetus for its use. If this course of action had not been taken we may well have had to improve our overview of particular problems, develop better approximations, draw better analogies, *etc.* From our point of view, computation provides a viable approach to solving a problem; the more power a computer can give you, the more flexibility you have in obtaining a realistic solution.

We will outline below what we believe can be gained by using computer graphics and animation to examine the data produced by computers. Our main interest and motivation is not computing *per se*, but to study how atoms coalesce to form crystals in the growth of semiconductors and other materials by a process known as molecular-beam epitaxy (MBE). The examples shown will focus on simulations of homoepitaxial growth on Si(001).

6.2 MOLECULAR-BEAM EPITAXY

MBE is a thermal evaporation process in which neutral atomic and molecular beams are directed towards a heated substrate in an ultra-high vacuum environment [1,2]. The vacuum is needed to prevent any foreign atoms or molecules contaminating the surface, becoming embedded in the growing crystal and diminishing the purity of the crystal. The term "molecular" refers to the fact that there are neither collisions nor chemical reactions among the particles in the beam prior to arrival at the substrate; in other words, the deposition is ballistic. The particles in the beam are usually either atoms or small molecules. In most circumstances, the evaporated species adhere to the target surface which, combined with the relatively slow growth rate (typically, approximately 1 atomic layer per second), yields enough control over the amount and type of material deposited to allow changes of compositions across an atomic layer. In this way, heterogeneous materials with atomically-abrupt interfaces can be fabricated.

A schematic illustration of the MBE apparatus is shown in Fig. 1. The substrate is mounted on a platform over which are located the Knudsen cells, one for each type of material. These cells are crucibles containing material kept at a temperature high enough to evaporate the material and so create the molecular beams. The growth of the material is monitored *in situ* by reflection high-energy electron-diffraction (RHEED), which requires an electron source, or "gun," and a screen for measuring the electrons reflected off the surface. The mounting platform can be rotated during growth so that the effects of inhomogeneities in the molecular beam are minimized. Some method of heating the substrate is essential to promote the mobility of atoms on the surface, which smoothens the growing surface.

Much of the excitement about the materials grown by MBE stems from the behaviour of electrons in the presence of structures with characteristic

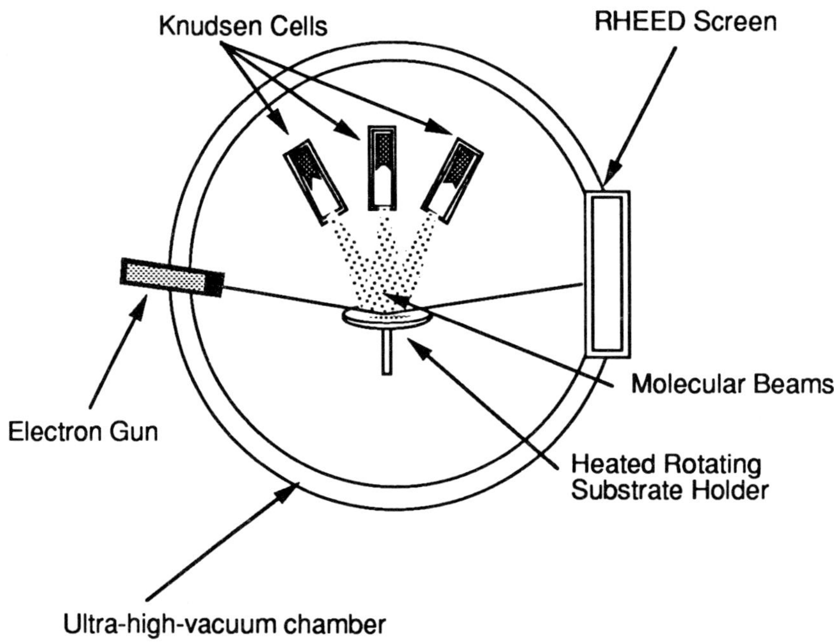

Figure 6.1 A schematic illustration of the experimental arrangement for molecular-beam epitaxy.

dimensions comparable to the de Broglie wavelength of the mobile electrons, and the ability to tailor the spatial and electronic properties of these structures [2]. Examples include confinement of electrons in a quantum well, mini-band formation in a quantum-well superlattice, and resonant tunnelling. The electronic and optical properties of these materials may, to a certain extent, be engineered by using selected combinations of materials, *e.g.*, GaAs with either AlAs or AlGaAs, or Si with Ge, and choosing the appropriate spatial modulations of the composition.

Many of the properties of materials produced by MBE are influenced by structures and interactions occurring on the scale of atomic dimensions. Theoretical models and simulations are useful for addressing the fabrication of such "atomically-engineered" materials by providing a window to the microscopic movement of atoms to highlight the important steps in the growth process. Since the atomic-scale characterization of materials grown by MBE is usually carried out upon cessation of growth, computer modelling may be the only way to observe the atom-by-atom formation of the material. Furthermore, the sub-monolayer control of the deposition in MBE has encouraged the fabrication of structures with progressively smaller dimensions, which in turn has necessitated closer examination of the growth process. When the lateral dimensions of the structures are large on atomic scale [approximately 1000Å, or approximately 250 atomic spacing on a GaAs(001) surface], the details of the growth process are not a limiting factor. However, in recent years, with lateral dimensions of heterostructures approaching a few tens of Ångstroms [4], atomic-scale imperfections that occur during the growth can severely undermine the integrity of the interface of two materials. It is in this arena that computer modelling is particularly useful.

6.3 MODELLING OF MBE

MBE is a process driven by kinetic phenomena on length scales comparable to the size of atoms and over times that are typical of those required to form and break chemical bonds. The current inability to probe into the details of events on such scales during growth is one of the primary motivations behind development of computer simulations. The two principal techniques that have been utilized are molecular dynamics [5-7] and Monte Carlo simulations [8-10]. These two methods are reviewed below.

6.3.1 Molecular Dynamics

Molecular dynamics proceeds by solving the equations of motion for every atom in the system, then using the calculated forces to determine the direction of movement of the atoms. All of the physics is contained in the forces acting upon each atom in the system, which are determined by the interatomic potentials for each atom in the system; this is the essential computational and theoretical ingredient of molecular dynamics simulations.

The equations of motion are solved for a sequence of time steps which, to ensure that the atoms do not diverge too far from their actual trajectories due to this time discretisation, must be less than the atomic vibrational frequency (approximately 10^{-13}s). As can be seen, there has to be an enormous number of computations performed before an appreciable time interval has elapsed. It is also important to keep in mind that at each stage the calculations are quite complex. Working out the competing forces of attraction and repulsion between all of the atoms in the system at a given instant requires a significant amount of computer time even when there are only a few hundred atoms.

6.3.2 Monte Carlo Simulations

Monte Carlo simulations of crystal growth make use of the fact that atoms spend the majority of their time in a complex but closed orbit close to their lattice positions. This can be seen in the molecular dynamics simulations shown in Figure 5.1 and Figure 5.3. Only very infrequently on the time scales considered by molecular dynamics will an atom change its average position from one lattice point to another. This involves confining the possible atomic positions to the lattice points which, in the case of crystal structures with few or no defects, is a sensible approximation.

The most commonly used lattice model is the so-called solid-on-solid (SOS) model [8,10,11], whereby the atoms within a given layer are supported directly by atoms in the underlying layer. In other words, unoccupied lattice sites (vacancies) are not allowed. Each surface configuration may then be described by an array **H** whose entries $h_{i,j}$ contain the numbers of atoms in the columns of atoms extending perpendicular to the surface. Kinetic processes such as adsorption, migration, and evaporation that characterize surface dynamics are specified in terms of rules for the addition or subtraction of atoms in the columns. Once the rules are specified, the evolution of the surface can be described by an equation that contains the transition rates among different states (the Master equation).

The Monte Carlo model replaces the detailed but unpredictable trajectory of an atom, with a probability of moving from one lattice point to another. All but the essential rate-determining processes are discarded, but the influence of these fast processes is retained in the randomisation of the processes that are treated explicitly. The occurrence of an event is determined by calculating or estimating the probability per unit time for the event, which is then compared with a random number generated from a uniform distribution between 0 and 1. If this random number is less than or equal to this probability, the event takes place, otherwise not.

The relevant time scale between events is the fastest rate process the system can generate, which is typically in the range 10^{-4}–10^{-6}s. Thus, for each time step in a Monte Carlo simulation, molecular dynamics will have to perform of the order of 10^8 steps, each of which takes a much longer time to evaluate than a Monte Carlo step. Considering that the number of time steps is so significantly reduced and the computation between each time step is also drastically simplified, the advantages gained from a Monte Carlo approach are evident.

Thus, the algorithm for generating the dynamics in this type of Monte Carlo simulation is very different from the Metropolis algorithm used in studies of thermodynamic equilibrium [12,13]. Among the many important differences that arises between the two algorithms is the question of the "time." In the algorithm describe here, the simulation time corresponds to the laboratory time, but in the Metropolis algorithm, there is no simple,

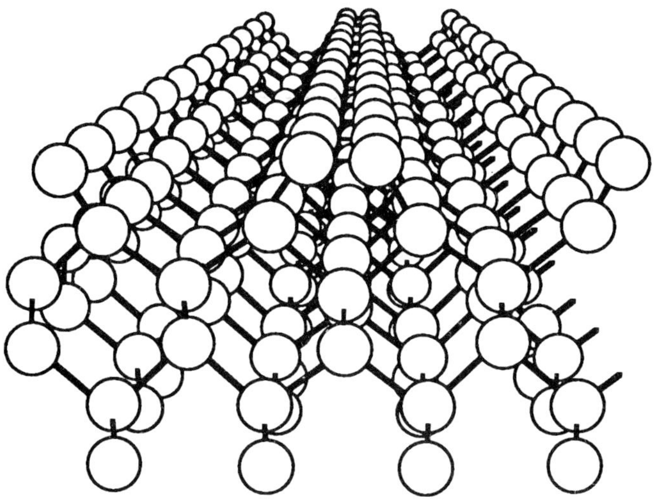

Figure 6.2 Diagram of the Si(001) surface viewed along the dimer rows.

direct relationship between simulation and laboratory times [14].

6.3.3 Model for Si(001) Homoepitaxy

The basic feature underlying many aspects of Si(001) homoepitaxy and heteroepitaxy is the formation of π–bonds between adjacent surface atoms to reduce the surface free energy of the bulk–terminated structure [15]. Scanning tunnelling microscopy (STM) of Si(001)-(2×1) surfaces [16,17] has shown directly that the dimers align in rows with a very low misalignment density giving the surface a corrugated appearance (Figure 6.2), but has also revealed a relatively large number of missing dimer defects, as well the presence of both buckled and non–buckled dimers in approximately equal numbers. Similar conclusions have also been reached by *in situ* STM studies of Si(001) grown by MBE [18,19].

We have chosen to retain the structural simplicity of the original SOS model by not explicitly including the dimers. Thus, fluctuations in the formation and dissociation of dimer pairs are neglected, but instead are included on average. The influence of dimerisation upon growth is included in the attachment of atoms to clusters, which is determined only by the numbers of nearest neighbours parallel to and perpendicular to the "dimer bond" axis. This leads to unphysical cluster configurations only in the early stages of growth; once clusters are formed, subsequent capture of atoms is correctly described by our rules, as evidenced by direct comparisons with STM images [18,19]. The primary structural constraint imposed by the Si(001) structure is that any feature associated with this surface dimer bonding should reorientate through 90° each successive monolayer.

In our implementation of the SOS model for GaAs and Si homoepitaxy, two surface processes are included, deposition and migration, and kinetic activity is restricted to surface atoms. Evaporation of surface atoms is usually neglected, as at typical growth temperatures negligible desorption flux is observed. Growth is initiated by the random deposition of atoms onto the substrate. The surface migration kinetics are then modelled by prescribing an isotopic inter-site hopping probability with an Arrhenius form: $k(E,T) = k_0 \exp(-E/k_B T)$ in which k_0 is a surface vibrational term, k_B is Boltzmann's constant, T is the substrate temperature, and E is the barrier to migration.

A surface atom has a configurational diffusion barrier E comprising a substrate term and a term representing the number of nearest-neighbour bonds formed parallel to the substrate: $E = E_S + n_\perp E_\perp + n_\parallel E_\parallel$ where E_S is the substrate term, $E_\perp$ is the nearest-neighbour barrier in the direction perpendicular to of the dangling bond of the surface atom, $n_\perp$ represents the number of bonds in that direction, with $E_\parallel$ and $n_\parallel$ being the same quantities in the parallel direction.

The strain induced by the reconstruction creates anisotropic surface bonds that are much stronger along a corrugation, or perpendicular to the direction of dimerisation, than along the orthogonal direction. For Si(001) we set $E_\parallel = E_\perp/10$ and, as the direction of the dangling bond rotates through 90° on successive layers, the direction of the barrier anisotropy changes accordingly. The justification for the anisotropy comes from step-energy calculations performed by Chadi [20] and Aspnes and Ihm [21], where it was found that steps that form parallel to the direction of the dangling bond are energetically less stable than the perpendicular steps. The parameters used in all the animations are: $E_S = 1.3$eV, $E_\perp + E_\parallel = 0.5$eV and $k_0 = 10^{13}$ s^{-1}.

The random nature of the kinetic processes enters both in the deposition stage, where the site at which an atom arrives is chosen at random, and in the migration of atoms along the substrate, where the probability of hopping from one site to a nearest-neighbour site is determined by the number of bonds the atom has formed; the more bonds formed, the less mobile the atom. The level of detail can be increased to whatever degree desired, though the simplicity of our model in conjunction with fast algorithms has enabled simulations to be carried out using realistic substrate temperatures and growth rates (typically 1 monolayer/s) which can be maintained for extended periods of time (up to 1500 monolayers) on large lattices (up to 500×500 units). Molecular dynamics simulations of MBE, on the other hand, have been confined to small lattices for short growth periods at unrealistically high growth rates. Furthermore, we have been able to perform the vast majority of our simulations on Macintosh IIci desktop computers!

6.3.4 Epitaxial Growth on Vicinal Si(001)

An important feature of growth on Si(001) is the strong dependence of the growth characteristics upon the method of preparation used on the substrate. More precisely the growth mode depends upon the temperature and duration of substrate annealing prior to growth. This dependence can be explained in terms of defects in the substrate in the form of surface steps.

A real crystal surface is never precisely flat because it is impossible to align the surface exactly along a crystal axis. Of course, the more accurate the correspondence between the surface and the crystal plane, the greater the distance between steps. In the context of growth, steps are defects in the surface which, depending upon the conditions of growth in relation to their separation, play an important part in determining the growth mode. In the case of the Si(001) surface the situation is complicated still further by the presence of the dimer structure. Due to the fact that the orientation of surface dimers changes on each successive layers, the direction of dimerisation changes across a monatomic step (Figure 6.3). The orientations of steps, like that of a surface, can never be precise, but can be arranged so that on average most of the step follows a line defined by the surface atoms, *i.e.* along a surface lattice vector. Due to the change in direction of the dimers, the surface produces two distinct types of step: one with the dimers on the terrace above the step, perpendicular to the edge and one where the dimers are parallel to the edge. We will refer to these two types as S_A and S_B respectively, with the same nomenclature for the terrace behind each step (Figure 6.3).

The dimer bond-axis on the reconstructed surface defines the direction of the strongest bonding. This means that the orientation of the dimers at the step edge affects the behaviour of the steps. For instance the S_B step is less energetically stable than the S_A step and so is more likely to roughen, thereby introducing a large number of kink sites. Also, the presence of the bond anisotropy, in conjunction with monatomic steps, affects what can be observed. As the anisotropy on a type-B terrace is orthogonal to that on a type-A terrace, the growth modes on the two types of surfaces are rotated relative to one another. A surface that is comprised of type-A and type-B terraces will on average appear quite symmetrical. Thus, for any averaged quantity to have a directional dependence the surface must comprise a majority of one type of terrace. Such single-domain surfaces can be formed by forcing the steps to be biatomic in height. A schematic representation of such steps can be found in Figure 6.3. These biatomic

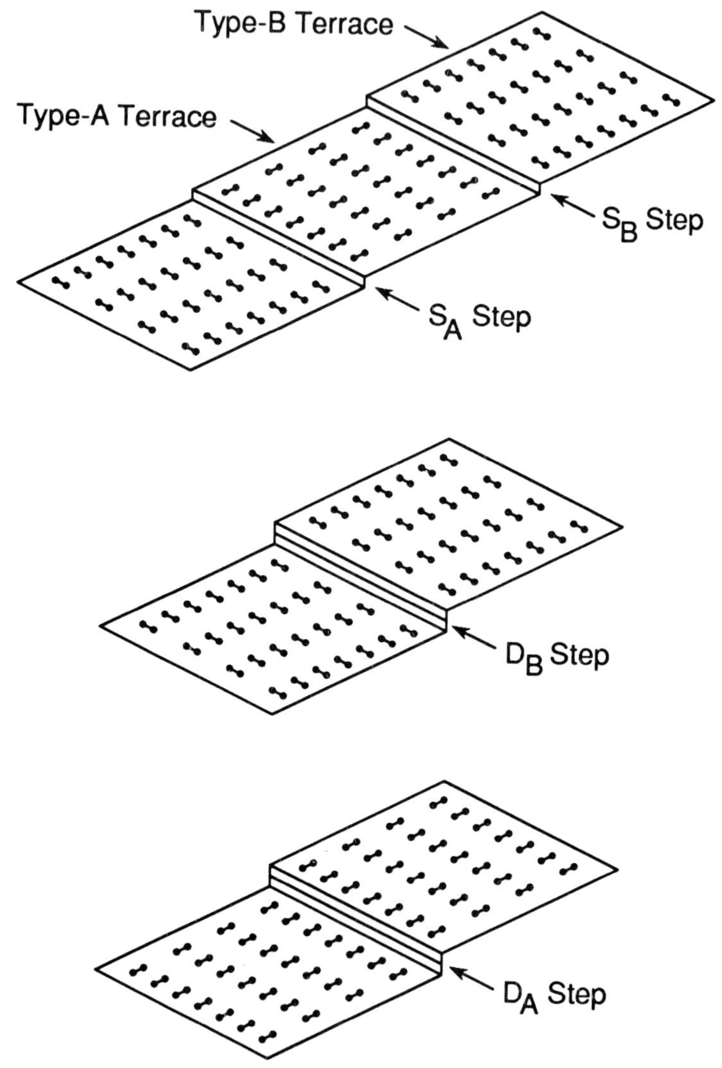

Figure 6.3. Schematic illustration of the notational convention for referring to single (top panel) and double steps (bottom two panels) and the associated terraces. The dimers are represented by filled dark circles connected by lines.

steps are label in the same way the monatomic steps are, but replacing the "S" with a "D."

The substrate can be forced to form a single domain with biatomic steps by the careful use of certain surface preparation methods. Calculations have shown that the D_B steps have an energy of formation that is lower than the sum of the energies of formation of the S_A and S_B steps [20]. This means by annealing the surface at sufficient temperature (approximately 1000°C) for extended periods (approximately 1 hour), pairs of monatomic steps can be made to form biatomic steps. Once this has been achieved the surface is made up of a single domain, constructed from type-B terraces. Such effects are demonstrated by the fact that surfaces subjected to such annealing, exhibit oscillations in the RHEED specular intensity, exhibiting either a monolayer or bilayer period, depending upon the azimuthal orientation of the electron beam [22]. However, without being annealed, *i.e.* composed of mixed domains, there can only be observed a monolayer period in the RHEED intensity oscillations, independent of the azimuth [23]. Notably the bilayer oscillations are considerably more sustained than the monolayer oscillations.

6.4 ANIMATION

Why is it necessary to consider such a complex task as animating a Monte Carlo simulation? In most cases, observables and other diagnostic parameters, typically averages of some kind such as the coverage of different layers, are calculated during the course of a simulation. Once the simulation has been carried out, or even while it is still progressing, these quantities can be plotted and analyzed. However, the identification of meaningful quantities is not always a straightforward task, in addition to which the use of average values can be misleading. An average value that is difficult to interpret may be driven by temporal fluctuations or, more likely, stabilised by a generated structure or correlation that the averaging procedure cannot detect. The use of many distinct observables can improve things considerably, but this leads again to a large and unmanageable amount of data in a form that is difficult to interpret. If a particular variable is being used as a comparison to theoretical or experimental results, then anomalous (or apparently anomalous) behaviour would have to be examined by an independent and more detailed study of the system.

It is when the chosen observables break down, or need a more detailed interpretation, that we require a method of monitoring as much of the data produced as possible. These data should be presented in a form that is accessible, preferably to an untrained observer. For this reason, the ability we all posses to monitor the time development of the world around us and assimilate huge amounts of visual data from which we can recognize spatial and temporal correlations, presents an ideal means of introducing new observational techniques. If the simulation data is translated into a form that makes use of this existing capability, *i.e.* animation, we have an entirely new way of "looking" into the microscopic world of atoms.

6.4.1 The Problems We Face

Any attempt to understand the growth kinetics of MBE must include atomic-scale interactions and yet describe structures up to approximately a micron in size. The structures of experimental interest range from the atomic scale to the collective behaviour of many interacting atoms that form mesoscopic structures. Thus, we are dealing with a problem that is concerned with the interaction of atoms on a scale where individual atoms can in some sense be resolved, but in sufficient numbers to simulate the structures that can be grown. An additional factor to consider is the time over which the simulation is performed. The time to grow a monolayer of material in MBE can vary from seconds to minutes; this should be taken into account in the model used. The simulation can thus involve many thousands, and sometimes even millions, of atoms that must be followed over an extended period of time. The fact that so much information is generated, particularly as the data is produced in spatial *and* time dimensions, makes data analysis a major problem.

Regardless of which model or type of simulation that is used, there is a problem in the sheer enormity of the data produced. The position and type of every atom in the simulation can be stored at regular intervals, with the time interval being dictated by the resolution required to follow the fluctuations in the system. From the numbers we have quoted for the size of systems, it is apparent that data transfer rapidly becomes unmanageable. We can, of course, resort to the "snapshots" approach, but this becomes impractical, as the required number of snapshots increases.

In addition to this is the difficulty of following the time-development of large systems between snapshots. Particularly difficult to follow in any static rendering are the fluctuations in the system, for while the average behaviour of a system or a part of a system can be relatively easily characterized, the deviation from the average in very small regions are extremely important for atomic-scale structures. For a proper understanding of the growth behaviour these fluctuations need to be monitored.

6.4.2 The Problems We All Face

The problems we face are not unique to simulations of MBE. The explosion of activity in many other areas of computational physics and engineering has been accompanied by ever-expanding volumes of data being generated.

Fortunately, developments in computer graphics have complemented this expansion. The combination of graphics and complex data sources has led to the introduction of "data visualization" techniques for presenting large quantities of data in ways that can be assimilated more readily.

In our particular case, the problem of data visualization hinges upon the time dimension. Since we are concerned with a growing surface, it is relatively simple to produce a two-dimensional representation of an instantaneous configuration, or a "snapshot" of the simulation. But, while a single snapshot can characterize the system at a particular time, it may still remain difficult to ascertain the extent of activity within the system or to follow the trajectory of the system from one snap shot to the next. A more appropriate solution would be to watch the time evolution of the system. To do this and produce the results in a widely usable form we use video animation.

6.4.3 How to Animate

The details of animating a sequence of data are quite complex, although the general process is relatively straightforward. To generate the animations for our simulations of MBE, we used the WINSOM general-purpose graphic research software developed at the IBM UK Scientific Centre (which is described in Chapter 3) to generate raster images of solid models representing the surfaces.

The first task was to determine how best to represent "realistically" a growing simple-cubic surface. We found by experimentation with static images and by appealing to earlier work in the literature that a schematic representation based upon "cubic" atoms provided the simplest and in many ways the clearest representation of the model. For applications to MBE the distinction between rough and smooth surfaces and step edges is of paramount importance in assessing the characteristics of growth and for the potential performance of a device. Building a surface from contiguous cubes provides a more readily apparent discrimination between rough and smooth than a surface built from, say, spheres.

One raster image was generated for each unique frame of the video frame, there being many hundreds of such frames in the final animations. A typical image took about 10 minutes to generate, though for this particular application, specially-designed software could have taken advantage of features in the graphics, such as polygonal shading to substantially reduce the required computational resources. More details of the animation process for MBE may be found in Ref. [24].

The medium to be used presented its own problems. Since the animation is intended for as wide a distribution as possible, from conferences to classrooms, the most convenient display technology is domestic television.

This however, introduced constraints upon the quality of graphics achievable, most evidently in terms of the spatial and colour resolution of the equipment. In addition, encoding the graphics into video signals imposed further limitations. Colours for normal graphics can be quite vivid, but must be toned down for television, and any sharp spatial modulation of colours which would otherwise be severely distorted, must be avoided.

One of the problems that can be encountered in resolving spatial detail is known as aliasing, for example, when a diagonal line appears as a "staircase." Fortunately this effect is less noticeable on television displays, and it was not necessary to employ "anti-aliasing" techniques (by artificially "smearing" the lines). However, the low chromatic resolution necessitated the introduction of additional edge lines to reinforce the detail. Such factors are important for the animation of MBE simulations: different shades and colours are used to differentiate among atomic layers and to distinguish between edges and terraces, while edge lines emphasize the abruptness between the edge and the terrace. In representing the Si(001) surface, it was necessary to distinguish between the two types of domains. This was done most effectively by using a different colour for each domain. It is evident from Figures 6.4 and 6.5 that the surface is constructed of two domains and each can easily be distinguished.

The final version of the videos show each simulation animated at three different rates. The fast rate illustrates the kinetics processes in real time. A real-time animation rate is useful for conveying the extent of kinetic activity during growth and during the equilibration of the surface, but cannot be used to focus upon individual processes. The medium rate is that most suitable for showing to audiences in seminars, where attention can be drawn to individual processes. The slowest rate is valuable for close scrutiny of the substrate.

6.5 ANIMATIONS OF Si(001) HOMOEPITAXY

All of the general points regarding the advantage of animation mentioned previously appear in one form or another when applied to our MBE simulations. The advantages stem from the level of detail concerning the kinetic activity contained in the animation. While an average description of the process can be easily obtained with quantities such as the density of surface steps or intensities in the diffraction pattern (observables that describe the system instantaneously by a single value), the statistical fluctuations are much more difficult to address theoretically. Within the simulations, guided only by the temporal behaviour of averages and snapshots, identifying the important fluctuations and meaningful averages becomes an extremely difficult task.

The use of animation to visualise the behaviour of the growing system as a whole has been particularly informative in the case of growth dominated by the presence of steps on the vicinal surface. The details of step profile fluctuations, which were not apparent in averaged or static analyses of the simulations [25], are shown in full dynamic splendour. Stills from the video demonstrate the irregularity of the step structure, but can not convey how it is likely to alter with time, or how one generated structure is likely to lead to another.

6.5.1 The Animated System

The animations have proved particularly useful in adding to our understanding of Si(001) homoepitaxy. We chose to animate a system with a substrate size of 120×120. This increased the degree of realism in comparison with the 40×40 lattices used in earlier animations [24], a consideration that cannot be stressed too strongly. It is all too easy to construct correlations that do not exist, particularly if the system size is small. What appears to be a large fluctuation in a small system may in fact contribute very little if the scale is increased, and so is attributed with more significance than it actually deserves. This is quite apart from the fact that increasing the size reduces the effect of the boundary of the system. The main disadvantage of the larger systems from our current stand-point, is the drastic increase in the graphical processing requirements [24].

We have mentioned that surface steps often play an important role in growth by MBE, depending upon the growth conditions, such as substrate temperature, the average distance between steps and the rate of growth. If the conditions are such that an atom deposited on the surface is more likely to encounter another diffusing atom before it reaches a step, then growth will be initiated at such meeting points, or nucleation sites. This is not quite a precise description, as in reality it also depends on the time such nucleation sites exist. Growth that proceeds by the nucleation and growth of clusters of atoms tends first to roughen the surface, as the clusters begin to grow, then to smoothen, as the clusters coalesce. Such behaviour is easily identified by oscillations in some average quantities with as period equal to the time required to deposit one monolayer. However, if the substrate temperature is high enough, adatoms are mobile enough to reach the step edge before encountering another atom. In this case, the steps dominate the way the surface grows. As more atoms are captured by the steps, the step edge advances or "flows." Under such conditions the average surface morphology is constant in time. If any more detailed information is required, then the nature of the fluctuations has to be considered.

6.5.2 What Can Be Gained

Our intention is to illustrate that even in the case of our simple model of Si(001) homoepitaxy, the behaviour exhibited is extremely complex. The details of the system can be found by prolonged analysis of the data produced, but as has been pointed out, this quickly becomes an impossible task. Animation provides an efficient method of observing the evolution of the entire system, including its fluctuations, but does not necessarily need to give quantitative results. What we require is a method of guidance, something that will give us a new perspective and allow us to identify new observed quantities, which we can then use as a basis of quantitative interpretation.

The animation methods used have in general led to a better understanding of the system we are dealing with. It is impossible to outline all the benefits we have gained, as some have been extremely indirect. For example, in watching the dynamics of the interplay between the two domains, it became clear that this was the driving force in the characteristics features of growth of Si(001). As an example of an indirect benefit, insights into the behaviour of the model have been used as a basis for an analytic approach to growth in the step flow mode [26,27]. Much has been learned along the way about fields outside our main discipline, and it has been just plain fun.

6.5.3 Behaviour during Growth

Even when considering growth on Si(001) that is dominated by the formation, growth and coalescence of islands on the surface, the visualisation method proved extremely useful. Although in this case the growth was well-understood from an analysis of averaged quantities, a demonstration of the actual growth sequence gives a dramatic clarification of the cooperative mechanism driving the system. What the animation shows clearly and quite surprisingly, is the degree to which a domain must complete before the next layer is formed. Due to the anisotropy in the barrier to hopping discussed in Section 6.3.3, the growing clusters are elongated, and the direction of elongation is rotated from layer to layer. This means that a layer must be almost complete before the next one can begin to form. Although this fact was known from the anisotropy in electron diffraction patterns, without seeing the highly cooperative way that the clusters in a given layer coalesce, it is difficult to have a genuine appreciation of just how much of the layer needs to form before the next one begins. The aspect ratio of a growing cluster does not change drastically while they are growing, but when the clusters begin to coalesce, they form extremely long thin troughs in the surface. These troughs are a barrier to clusters forming in the next layer and are a persistent feature. Since the width of these surface holes is small, adatoms tend to deposit at the walls of the hole rather than form nucleation centres within it, so they tend to fill by means of the two edges propagating towards one another. As a consequence they are likely to last until the layer has almost completed.

Figures 6.4 and 6.5 show the morphology of the surface after 1.5 and 2.5 monolayer deposition, on single-domain and double-domain surfaces, respectively. They show quite clearly the change in the direction of cluster elongation, but fail to convey the cooperative way surface troughs prevent the formation of the next layer. This cooperative behaviour is the bases of the stability of the bilayer oscillations in the RHEED intensity, which is still present within the simulation after 1000 monolayers have been deposited.

The animation proved most valuable where the growth proceeded by the advancement or flow of steps. We have already stated that monatomic steps on a vicinal Si (001) alternate in stability. As such, one of the step types has a much lower energy of formation and is therefore more susceptible to thermal break-up. This means that every alternate step on the surface has a much rougher profile than either one of its relatively stable neighbouring steps. The consequence of this is that the surface has an alternate rough/smooth profile of the steps which has been confirmed by scanning tunnelling microscopy [18,19]. Under the non-equilibrium conditions of growth, the less stable step advances (Figure 6.6), so that the predominant amount of the surface is terminated with an unstable step. The reason for this advancement had again been explained by careful consideration of the behaviour of average quantities and surface snapshots. Although the step is the least stable, it also presents it strongest bond to any adatom that finds itself in an adjacent site. From this we know that the residence time for such adatoms is longer than anywhere else on the surface. Whilst material is being added, such atoms are the most likely to be trapped by incoming material and hence the step becomes the fastest region of growth. It does not advance up to the step in front, because if it did material would add only to the bottom step and so separate them.

6.5.4 Attainment of Equilibrium

Something that could not be explained before the animation was the behaviour of the surface under relaxation. After the unstable step has been allowed to dominate the surface, the flux is stopped and the surface is allowed to relax. The step recedes, so that there is approximately an equal amount of each type of surface. Experimentally, this is observed by monitoring the half-order beams, whose intensity is a measure of the relative coverages of the two types of terrace [28]. Although the average measure of the surface demonstrates the effect, it does not explain its origin. From the animation it is quite apparent that at the temperatures needed for this type of growth, without the stabilising effect of adding material, the step is extremely unstable. In fact the position of a particular point on the step fluctuates rapidly around its average. The magnitude of the fluctuations exceed the inter-terrace spacing so they are bounded by the stable steps and hence its optimum average position is half-way between them.

The same effect is also partly responsible for maintaining a separation between the two types of step during growth. The fluctuations of the step position are much smaller during growth, so the two steps can approach each other. However, there are still some fluctuations and for the same reason as in the equilibrium case the steps are kept apart. It can now be realised that the effect of domain dominance is dependent upon the size of the terrace relative to the fluctuations, or temperature. If the terraces are large or the fluctuations are small the surface will not recover to equal domain coverage, an effect that is observed experimentally. Such an effect would be difficult if not impossible to identify without such a continuous way of observing temporal behaviour.

6.5 CONCLUSIONS

The ability to view the temporal evolution of individual atomic configurations has significantly added to our understanding of the solid-on-solid model and to the microscopic growth kinetics of MBE. Particularly fruitful have been the applications to GaAs(001) and Si(001), which have reproduced experimentally observed growth modes and highlighted the dynamic nature of terraced surfaces at higher temperatures. The development of visualization techniques such as those described here hold promise of an exciting future in the presentation and analysis of scientific information, including possible applications for emerging computer technologies, which take advantage of the inherent concurrency of performing and animating a simulation.

ACKNOWLEDGEMENTS

We are extremely grateful to Mr. M.W. Ricketts and his colleagues at the IBM UK Scientific Centre for their production of several video animations. S.C. thanks the U.K. Science and Engineering Research Council for support.

REFERENCES

1. A.Y. Cho, "Growth of III-V semiconductors by molecular beam epitaxy and their properties," *Thin Solid Films* **100**, 291 (1983).
2. B.A. Joyce, "Molecular beam epitaxy," *Rep. Prog. Phys.* **48**, 1595 (1985).

3. M.J. Kelley and R.J. Nicholas, " The physics of quantum well structures," *Rep. Prog. Phys.* **48**, 1699.
4. J.M. Gaines, P.M. Petroff, H. Kroemer, R.J. Simes, R.S. Geels, and J.H. English, "Molecular-beam epitaxy growth of tilted GaAs/AlAs superlattices by deposition of fractional monolayers on vicinal (001) substrates," *J. Vac. Sci. Technol. B* **6**, 1378 (1988).
5. D.W. Brenner and B.J. Garrison, "Microscopic mechanisms of reactions associated with silicon MBE: a molecular dynamics investigation," *Surf. Sci.* **198**, 151 (1988).
6. M. Schneider, I.K. Schuller and A. Rahman, "Role of relaxation in epitaxial growth: a molecular-dynamics study," *Phys. Rev. Lett.* **55**, 604 (1985).
7. M. Schneider, I.K. Schuller and A. Rahman, "Epitaxial growth of silicon: a molecular-dynamics simulation," *Phys. Rev. B* **36**, 1340 (1987).
8. J.D. Weeks and G.H. Gilmer, "Dynamics of crystal growth," *Adv. Chem. Phys.* **40**, 157 (1979).
9. A. Madhukar and S.V. Ghaisas, "The nature of molecular beam epitaxial growth examined via computer simulation," *CRC Crit. Rev. Sol. St. Mater. Sci.* **14**, 1 (1988).
10. D.D. Vvedensky, S. Clarke, K.J. Hugill, A.K. Myers-Beaghton and M.R. Wilby, "Growth kinetics on vicinal (001) surfaces: the solid-on-solid model of molecular-beam epitaxy," in *Kinetics of Ordering and Growth at Surfaces*, M.G. Lagally, ed. (New York, Plenum, 1990), pp. 297–311.
11. T.L. Hill, "Statistical mechanics of multimolecular adsorption III. Introductory treatment of horizontal interactions. Capillary condensation and hysteresis," *J. Chem. Phys.* **15**, 761 (1947).
12. K. Binder (editor), **Monte Carlo Methods in Statistical Physics.** Topics in Current Physics **7** (Berlin, Springer, 1986)
13. K. Binder (editor), **Application of the Monte Carlo Method in Statistical Physics.** Topics in Current Physics **36** (Berlin, Springer, 1987).
14. H.C. Kang and W.H. Weinberg, "Dynamic Monte Carlo with a proper energy barrier: surface diffusion and two-dimensional domain ordering," *J. Chem. Phys.* **90**, 2824 (1989).
15. J.D. Levine, "Structural and electronic model of negative electron affinity on the Si/Cs/O surface," *Surf. Sci.* **34**, 90 (1973).
16. R.M. Tromp, R.J. Hamers and J.E. Demuth, "Si(001) dimer structure observed with scanning tunneling microscopy," *Phys. Rev. Lett.* **55**, 1303 (1985).
17. R.J. Hamers, R.M. Tromp and J.E. Demuth, "Scanning tunneling microscopy of Si(001)," *Phys. Rev. B* **34**, 5343 (1986).
18. A.J. Hoeven, J.M. Lenssinck, D. Dijkkamp, E.J. van Loenen and J. Dieleman, "Scanning-tunneling-microscopy study of single-domain

Si(001) surfaces grown by molecular-beam epitaxy," *Phys. Rev. Lett.* **63**, 1830 (1989).

19. Y.-W. Mo, B.S. Swartzentruber, R. Kariotis, M.B. Webb and M.G. Lagally, "Growth and equilibrium structures in the epitaxy of Si on Si(001)," *Phys. Rev. Lett.* **63**, 2393 (1989).
20. D.J. Chadi, "Stabilities of single-layer and bilayer steps on Si(001) surfaces," *Phys. Rev. Lett.* **59**, 1691 (1987).
21. D.E. Aspnes and J. Ihm, "Biatomic steps in Si(001) silicon surfaces," *Phys. Rev. Lett.* **57**, 3054 (1986).
22. T. Sakamoto, N.J. Kawai, T. Nakagawa, K. Ohta and T. Kojima, "Intensity oscillations of reflection high-energy electron diffraction during silicon molecular beam epitaxial growth," *Appl. Phys. Lett.* **47**, 617 (1985).
23. J. Aarts, W.M. Gerits and P.K. Larsen, "Observations on intensity oscillations in reflection high-energy electron diffraction during epitaxial growth of Si(001) and Ge(001)," *Appl. Phys. Lett.* **48**, 931 (1986).
24. B. Ricketts, D.D. Vvedensky and S. Clarke, "Seeing is Believing," *Physics World* **2**(12), 39 (1989).
25. S. Clarke, M.R. Wilby, D.D. Vvedensky, T. Kawamura, K. Miki and H. Tokumoto, "Step stability, domain coverage, and nonequilibrium kinetics in Si(001) molecular-beam epitaxy," *Phys. Rev. B* **41**, 10198 (1990).
26. A.K. Myers-Beaghton and D.D. Vvedensky, "Nonlinear equation for diffusion and adatom interactions during growth on vicinal surfaces," *Phys. Rev. B* **42**, 5544 (1990).
27. A.K. Myers-Beaghton and M.R. Wilby, "The dynamic variations of terrace length during growth on stepped surfaces," *J. Phys. A* **24**, L35 (1991).
28. T. Sakamoto and G. Hashiguchi, "Si(001)-2×1 single-domain structure obtained by high-temperature annealing," *Jap. J. Appl. Phys.* **25**, L78 (1986).

7

Animation of Quantum Scattering Events using Hypercard

David M Halstead and Stephen Holloway

Animation has long been recognised as an essential tool in relating large amounts of information to an audience. Not only does animation utilise the sense of sight, which is accustomed to recognising patterns, but by altering the image progressively it also allows trends and behaviours to become evident in a natural way to the observer. The quantum scattering events presented here are ideally suited for animation, occurring in the time domain but displaying phenomena beyond the experience of the Newtonian world of our everyday life. The aim of this work is to illustrate an inexpensive way by which interactive animation may be achieved without the cost associated with workstations both financially and in graphics coding expertise. By use of a simple, standard, graphics oriented software engine available on all Macintosh computers, an entry level into animation may be obtained which should fulfil most basic requirements.

7.1 INTRODUCTION

The need to present the progression of time-dependent dynamics has been in existence since computers first came into use. An early landmark in the field of computer-generated graphics was made by Goldberg [1], who used an oscilloscope to generate individual frames that were then photographed and made into a series of short movies. At the time of Goldberg's work, it was prohibitively difficult to generate real-time movies on a computer, but with the advent of cheap, powerful, high resolution graphics machines, the whole process may now be done in-house by a desktop computer. Data rapidly displayed on a computer screen may also be directly captured onto video tape with acceptable results. The ideal rate of 50 frames a second is not practical on most mini-computers and would also require excessive calculation and storage of data, and would lead to only a marginal improvement in presentation clarity.

The choice to be made is whether to draw the image frame by frame, storing each as a raster file made up of pixels, or to draw the frames in real time and use a frame buffer to update the screen quickly. The problem with the first method is that it requires large amounts of storage if the image is large and, after rasterization, the viewpoint, perspective and scale cannot be altered. The advantage is that it gives a way of displaying images at a high speed which is independent of their complexity. If the image is generated as individual elements, it becomes trivial to alter the presentation interactively, but the speed of update will rise with image complexity. The approach to be presented here uses the first method, but with image compression to reduce the storage overhead required.

This chapter is organised as follows. Section 7.2 will outline the problem to be studied by dynamics, in a way that not only makes use of the images to help visualise important quantum effects, but that also has pedagogic value in illustrating it in a way other than as an abstract time-independent mathematical exercise. In the section 7.3, an illustrated example will be given on how the individual data frames so generated may be animated using HyperCard for the Apple Macintosh. Finally, section 7.4 will summarise the pros and cons of HyperCard animation and will outline some techniques which may be used to animate sequences of more complex images than those presented in the body of this work.

7.2 SOLUTION OF THE SCHRÖDINGER EQUATION

It has already been mentioned that one of the first exercises in computer generated movies involved the solution of the time-dependent Schrödinger equation (TDS). It therefore seems appropriate that this same problem should be now re-examined using new techniques which greatly simplify all aspects of the problem and should improve the accuracy, efficiency and the presentation quality of the scattering event simulations.

The problem of one-dimensional quantum scattering has long been used as a teaching tool whereby simple potentials such as a square barrier or well may be solved easily by matching wavefunctions and derivatives of wavefunctions at boundaries. More realistic potentials have also been solved, but with greater difficulty. For example, in the Eckart potential [2], a smooth barrier separates the initial and final states, which may be of different asymptotic potential energies. These problems were solved independent of time for a plane wave having, by definition, only one momentum component. In this work wavepackets will be used to simulate a plane wave which in turn gives an easy way of illustrating wave-particle duality, as well as the idea of scattering probability, tunnelling, interference effects and state projection.

For surface scattering problems, in order to reduce the vast number of degrees of freedom encountered, it is a common technique to reduce the coordinated treated to those directly associated with the process of interest. At the simplest level, direct reaction of an incoming species to produce products may be modelled as a restricted effective particle following an adiabatic ground state reaction path. A system exhibiting apparently direct activated adsorption is molecular hydrogen on copper. This system has provoked general interest both experimentally [3,4] and theoretically [5,6]. A compromise must be made between the accuracy of the calculation and its applicability to the system being modelled. To this end low-dimensional full quantum studies have been carried out for this system in the hope of understanding how the light mass effects may be manifested and tested for. To this end the time dependent Schrödinger equation is required to be solved for any proposed interaction potential energy surface (PES)

The time-dependent Schrödinger equation for the wavepacket ψ is

$$i\frac{\partial\psi}{\partial t} = (\mathbf{T}+\mathbf{V})\psi \tag{7.1}$$

where $\mathbf{T}$ is the kinetic energy operator and $\mathbf{V}$ is the potential energy operator. In writing (7.1), we have used units where $\hbar = 1$. The formal solution to (7.1) is

$$\psi(\Delta t) = \exp[-i\Delta t(\mathbf{T}+\mathbf{V})]\psi_0 \tag{7.2}$$

from which it is clear that the problem is one of evaluating the potential energy and kinetic energy operators to move the initial known wavefunction ψ_0 in time by Δt. If these are applied sequentially then errors due to their non- commutation result of the order $(\Delta t)^2$. If the kinetic operator is split symmetrically about the potential energy operator [7, 8], then the errors are reduced to $(\Delta t)^3$:

$$\begin{aligned}\psi(\Delta t) =& \exp[-i\Delta t\mathbf{T}/2]\exp[-i\Delta t\mathbf{V}/2] \\ &\times \exp[-i\Delta t\mathbf{T}/2]\psi(t) + O[(\Delta t)^3]\end{aligned} \tag{7.3}$$

In order to make the application of the operators as simple as possible, Fourier transforms are used to convert the complex real space wavefunction into its corresponding momentum space analogue. In momentum space, the kinetic energy operator becomes as simple to apply as the potential energy operator is in real space, i.e., by multiplying each meshed wavefunction value by a complex number which is constant with time but position-dependent. Figure 7.1 shows a flow diagram for the split operator method and illustrates how, by the combination of sequential half time step kinetic energy operators, a propagation loop may be set up requiring only two fast Fourier transforms (FFT) per iteration loop. The actual wavefunction is not calculated at each step, but as each complex number multiplication used here only alters the phase between real and imaginary components, the correct probability ($\psi^*\psi$) at each mesh point *is* generated during each time step. In most cases this probability will be sufficient to show the essence of the interaction dynamics, provided both the real and momentum space probabilities are shown. To illustrate this, the next section will give an example of a wavepacket scattering event using the split operator technique.

7.2.1 Dynamics through a Barrier

The use of square barriers has the advantage of greatly simplifying the task of analytical calculation of transmission coefficients, but gives results which are oscillatory, depending on the number of half wavelengths inside the barrier. If the barrier is smooth and so more representative of potential barriers occurring in reality, then the transmission probability rises smoothly from zero to unity with increasing particle energy. By use of the smooth barrier potential formulated by Eckart [2], an easily parameterised, flexible potential may now be employed having the form

$$V(x) = \frac{A\xi}{1+\xi} + \frac{B\xi}{(1+\xi)^2} \tag{7.4}$$

(with ξ is defined by $\exp(x/d)$), being zero at $x = -\infty$ and attaining an energy A at $x = \infty$. In order for there to be a barrier, $|A|$ must be less than $|B|$, resulting in a barrier of height $(A+B)^2/4B$ and width of $\sim 2\pi d$. For the purposes of this work A is set to zero which gives a single barrier of height $B/4$ which goes to zero away from the potential's centre. This simplifies the analytical solution for the transmission coefficient of a particle of mass m and energy E which may now be stated as

$$T = 1 - \frac{1 + \cosh(\pi d\sqrt{8Bm - 1})}{\cosh(4\pi d\sqrt{2Em}) + \cosh(\pi d\sqrt{8Bm - 1})} \tag{7.5}$$

This will be used to check the accuracy of the final state probabilities run using the potential meshed out onto a grid. This grid must be large enough to allow the interaction to finish without the edges being encountered, and fine enough to resolve the wavefunction (the maximum momentum is of the order $\pi/\Delta x$). The example used here will model a light particle [$m = 0.5$atomic mass units (amu), 911atomic units (au)] hitting a barrier

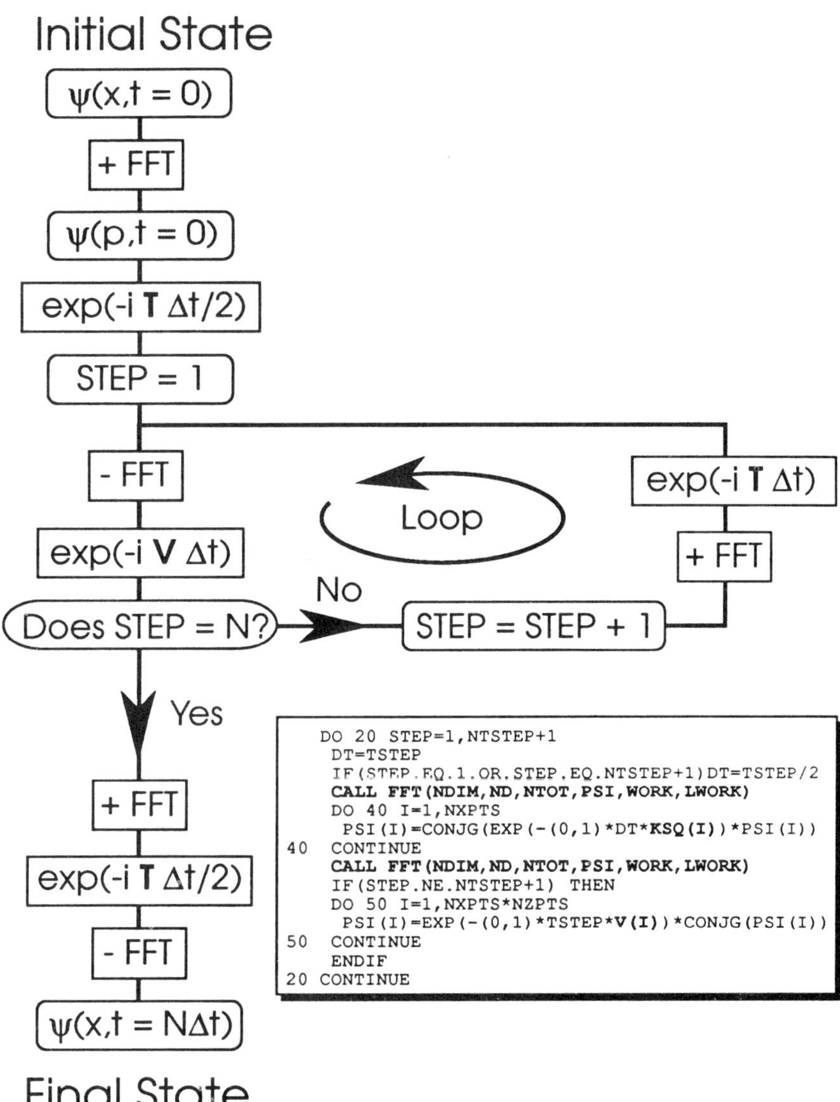

Figure 7.1. A flow diagram for the split operator algorithm showing how, inside the iteration loop, the two half time steps are combined to give an efficient cycle requiring two FFTs and two array multiplications. Also shown in the inset is the main loop of the code.

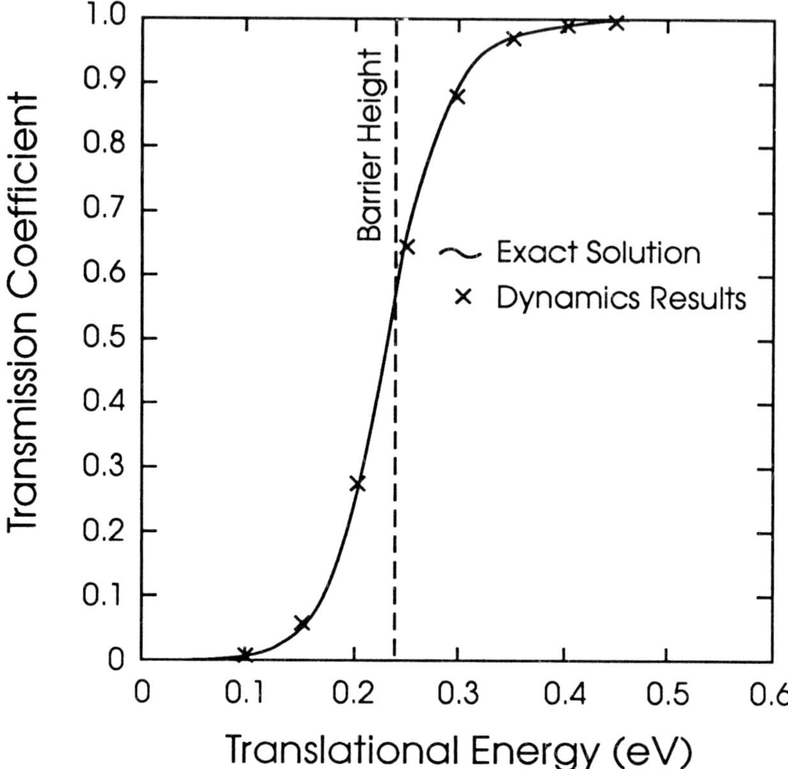

Figure 7.2. A comparison of the dynamical and analytical results for the variation of the transmission coefficients with energy for an Eckart barrier of height 240 meV.

of height 240meV and of width parameter $d = 1/\pi$ a.u. The resulting transmission coefficient is shown in Figure 7.2 for both the analytical solution, equation (7.5), and a series of dynamics runs in which the transmission coefficient is simply obtained by integrating over the wavefunction probability found beyond the barrier when the potential energy of the system indicates the interaction is complete.

The reason for this choice of "particle" mass is that it represents the reduced mass of a H_2 molecule while dissociating. The effects which the reduced mass has on the dynamics can be quite dramatic as it is only a quarter of the molecular mass for a homonuclear diatomic and so zero point energy effects as well as tunnelling probabilities become important. As can be seen in Figure 7.2 even at energies less than half the barrier height about 1% of the wavefunction is still able to tunnel through the barrier.

The initial wavefunction of the particle must be chosen in such a way that it simulates a plane wave of the required energy, yet is localised in space such that when the time development begins it has negligible overlap with the potential barrier. A flexible, yet simple form has been presented by Heller [9] which uses two complex parameters in addition to the mean position, x_0, and momentum, p_0, in its definition. The first of these two complex parameters, α, defines the width of the wavepacket in both real and momentum space, while the second, γ, provides for its normalisation and initial phase. The shape function chosen is a Gaussian, giving a wavefunction of the form

$$\psi(x, t_0) = \exp\left[i\alpha(x - x_0)^2 + ip_0(x - x_0) + i\gamma_t\right] \tag{7.6}$$

It is important to recognise that after the initial definition, no constrain is placed upon the shape of the wavefunction. If the width parameter α is defined to be purely imaginary, then the wavepacket will be of minimum uncertainty. This means that the product of the real and momentum space widths is unity (in atomic units) and will rise both forwards and backwards in time in the absence of a varying interaction potential. As the wavepacket has a localised width in real space, the momentum distribution is broadened away from the delta function obtained in the plane wave case. Thus the wavepacket may either be viewed as a localisation of a plane wave, or as a coherent summation of plane waves. By definition the initial momentum space wavefunction is also a Gaussian having the form

$$\psi(p, t_0) = \frac{1}{\sqrt{-2i\alpha_t}} \exp\left[i\frac{(p - p_0)^2}{4\alpha} + ipx_0 + i\gamma_t\right] \tag{7.7}$$

The sequence of frames from a typical scattering event is shown in Figure 7.3. Both real space and momentum space probabilities are shown, which in turn leads to a better understanding of the inter-relation of the two spaces not readily afforded by the analytical solution. In a typical run as few as 128 mesh points need be used to contain the interaction. A box length of 50au can safely resolve wavefunction energies of up to 0.6eV, but low energy runs may need more as wavepacket spread then becomes important as the finite width in momentum space produces broadening in real space for a packet in a potential free region of space.

The length of the time step required to give accurate propagation results depends on the rate at which the third order expansion to the exponential operator converges. As the problem is one of none commutation of the potential and kinetic operators it is only when ψ is finite in a region in which the potential energy is varying that errors accumulate. A rule of thumb is that the product $\Delta t E_{\mathrm{max}}$ must be less than 0.1, where E_{max} is the maximum energy encountered in the interaction. This gives a total

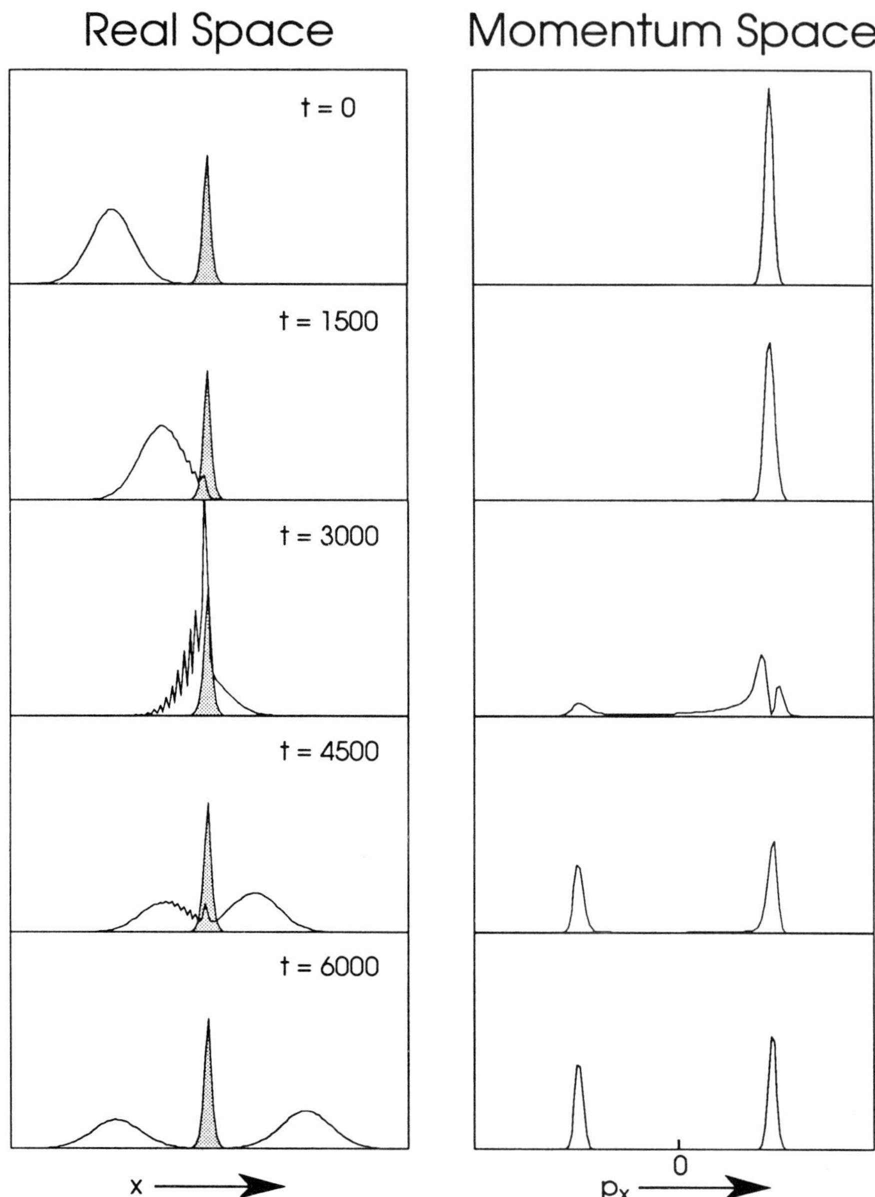

Figure 7.3. Five frames taken from the scattering of a wavepacket from an Eckart barrier of height 240 meV, which is shown shaded for clarity. The wavepacket energy is equal to the barrier height. The time of each frame is indicated in atomic units.

of between 200 and 500 iteration steps for this system depending on the kinetic energy of the particle. All that remains to be done is to extract the wavefunction probabilities at set intervals throughout the scattering event and then display them graphically. These images must then be animated by a process which will now be outlined.

7.3 STEPS TO GENERATING A HyperCard STACK

The basic feature employed in using HyperCard to animate pictorial data, is its ability to show a sequence of cards (the basic element of HyperCard). The ease with which cards may be linked and flashed to the screen as pixel images, linked to the automatic compression of the bit image data when it is stored, makes the manipulation and storage of images transparent once inside HyperCard. Given this, the major problem is to find the best route for obtaining the individual frames. Due to the simplicity of the graphics required to visualise a one-dimensional scattering problem, the number of possible ways to generate the line graphs and the ease with which the resulting images may be imported into HyperCard make it unnecessary to attempt to define an "ideal" way here.

Two different routes have been investigated, both of which give satisfactory result. The first uses a graphics terminal emulation application to communicate with a mainframe computer. As the simulation runs, Tektronix graphics information is sent to the screen at the required interval, for example every ten iteration steps. These images may then be copied into the clipboard, and pasted into HyperCard using techniques to be outlined below. The second way employs a simple graphing program of the sort widely available. These usually produce images consisting of straight line vectors as do the terminal applications graphics. When these are imported into HyperCard the image seen on the screen appears unaltered, but is now made up of a bit mapped copy of the original vector image.

The next section details the way in which some level of automation may be introduced. This is not only to simplify the process, but also to make it reproducible from frame to frame.

7.3.1 A Walk through an Example

In order to illustrate both the method of movie generation and its uses, an example will now be given, assuming that the image sequence to be animated may be obtained individually by copying from the application in which they were generated. It is helpful if the original image is of the correct size on the screen, as rescaling after importation into HyperCard is unsatisfactory as only a bit map image is stored. The first step is to have both the graphics source application and HyperCard open under MultiFinder. Although this is not essential (the Scrapbook may be used to store the series of images) it does help to automate the transfer process. Once the image is obtained on the screen, in this case from a graphics terminal emulation program, a macro recorder is started, then the image is selected and copied. The application is then switched to HyperCard from the apple menu and the image is pasted in. It is at this stage that the bit image mapping occurs. All that remains to do after this is to create a new card for the next frame and to return to the original application.

The macro recording must then be stopped, the next frame generated and the new transfer macro replayed. Experimentation may be required to get the image of the correct size and suitably positioned, but by using the macro facility, once obtained, the process may be exactly reproduced each time. The macro software used in this case was MacroMaker which is provided as a system utility with all Macintosh machines.

A typical movie so created may be between twenty and a hundred frames long. To see the movie immediately the text `show all cards` can be executed in the HyperCard message box, selected from the `Go` menu. This command is an example of HyperScript, the command language used to control HyperCard. There are many books on the subject of Scripting, but care must be taken as some books (including the manual provided with the machine) give only a cursory introduction to the Scripting command set. A good introduction to, and working knowledge of, HyperCard and HyperScript may be obtained from either of two introductory HyperCard books[10,11] and for the more advanced user there is The Tricks of the Hypercard Masters [12]. Although stack control via the message box this is quick and simple, it is better to set up a single mouse click to, for example, animate the stack, stop at the last card and wait to either repeat or branch to a related stack. This may be done by associating a short sequence of HyperScript commands with a card or a button.

For this work, only loops and links will be examined, these being sufficient to organise a series of inter-related movies and animate each in turn with a minimum of technical knowledge and intervention on the part of the end user. The animation HyperScript used to run a series of consecutive cards is shown below.

```
on mouseup
 repeat with count = 1 to number of cards-
1
  wait for 2 ticks
  go next card
 end repeat
end mouseup
```

The script is an interpreted sequence of commands which follow English grammar quite closely. In this case the script is entered into the **Script** section of the **card info** selection under the **Objects** menu. If this is done on the first card in the stack, then, after opening, the mouse button need merely be pressed anywhere on the card to initiate the animation sequence. The **on mouseup** to **end mouseup** bounds the commands to be executed when the mouse button is pressed. To stop the run at the last card, the repeat loop extends over one less than the number of cards in the stack, while the commands inside the loop control the animation speed and movement to the next card respectively. The script could just as easily have been allocated to a button in order to execute the loop commands. When the last card has been reached, it is a simple matter to redirect the stack back to the first card by using Script of the sort:

```
on mouseup
 go first card
end mouseup
```

or to make a button link in another stack or an index card by using the **LinkTo** option allocated to each button. Having created the stack it may be run free standing, or linked to from an index card which contains a button calling up its first card. After completion the last card may be linked back to the index also by using a button.

Embellishments may be made to the Stack such as resetting the Scripting level by the command **setUserLevel** n, were n defines the level of access to the user, 1 being the lowest and 5 the highest. In addition, if the commands:

```
on openstack
 hide menubar
end openstack
```

are associated with the Stack (place in the **Script** section of the **Stack Info** selection of the **Objects** menu) then the menubar at the top of the screen is removed. This not only serves to give slightly more space for images on the small screened Macintosh models, but also makes it harder for the inexperienced user to do any harm to the stack, as any changes made are stored continually unless the stack is locked. The menubar may, of course, be returned by executing the command **show menubar** at any point in the stack, or from the message window.

7.4 CLOSING REMARKS

We hope the above examples given some insight into the role that Hyper-Card animation may fulfil. There are certainly more powerful animation applications available for the Macintosh, but none can yet compete with its wide spread availability and ease of use. Below are listed some of the pros and cons of HyperCard animation:

Advantages

i. Uses a standard application which will run on all Macintosh systems
ii. Easy to interactively cross relate elements of several connected movies
iii. little expertise needed to create or use the stacks
iv. Hardware is available for interfaced to video cassette recorders
v. Speed of animation is little effected by the image complexity
vi. Spoken narrative may be included
vii. Easy to import images from other applications, or computers using standard formats, e.g., TIFF
viii. Stacks may be sent readily by electronic mail or by disk (a typical stack of 50 frames takes approximately 40 kB after compression.)

Disadvantages

i. Images are stored as bit images, so graphical information is mapped to the resolution of the screen
ii. Does not fully support colour
iii. Is not freely portable to other manufacturers machines
iv. Animation speed depends strongly on machine power which varies by an order of magnitude across the Macintosh product range.

Other consideration may come into play, such as the intended use of the stacks or the level of expertise of the end user. If the full benefit of their interactive nature is to be exploited then making a video of the screen display is self defeating. An ideal way of presentation to a large audience is the use of a liquid crystal overhead projector display. This gives a system that is portable, as a monitor is not required, but allows for random access of images and sequences.

The complexity of the images presented in this work has been minimal. This was intended, as the more complex the image, the more specialised is the nature of the users requirements. One way of importing large amounts of imaged data into HyperCard is by using a centralised computing facility to generate the images as vectors and objects, then to capture these from the terminal emulation application directly into HyperCard. This method has been used to generate stacks showing detailed contour maps of scattering simulations, giving complex animated images which are still manageable in size and quick to display and update once inside HyperCard. By use of these sequences, dynamic effects such as virtual excitations, resonance induced adsorption and vibrationally assisted dissociation have been illustrated to dominate processes which could otherwise have been overlooked if the final scattered state distributions where the only consideration.

Finally, we hope that the central two sections of this work can be perceived as free standing to some extent, in that quantum dynamics gives valuable insight as well as a way of solving the TDS in an efficient way, and a brief introduction to one of the many uses for HyperCard.

ACKNOWLEDGEMENTS

The authors would like to thank the organisers of the "Seeing is Believing" conference held in the Liverpool Surface Science IRC in the spring of 1990. In addition special thanks goes to Mick Hand for his early work in the use of HyperCard as an animation engine.

REFERENCES

1. A. Goldberg, H.M. Schey and J.L. Schwartz, "Computer-generated motion pictures of one-dimensional quantum mechanical transmission and reflection phenomena," *Am. J. Phys.* **35**, 177 (1967).
2. C. Eckart, "The penetration of a potential barrier by electrons," *Phys. Rev.* **35**, 1303 (1930).
3. G. Anger, A. Winkler and K.D. Rendulic, "Adsorption and desorption kinetics in the systems H_2/Cu(111), H_2/Cu(110), and H_2/Cu(100)," *Surf. Sci.* **220**, 1 (1989).
4. B.E. Hayden and C.L.A. Lamont, "Coupled translational-vibrational activation in dissociative hydrogen adsorption on Cu(110)," *Phys. Rev. Lett.* **63**, 1823 (1989).
5. D. Halstead and S. Holloway, "Quantum-mechanical scattering of H_2 from metal surfaces: diffraction and dissociative adsorption," *J. Chem. Phys.* **88**, 7197 (1988).
6. D. Halstead and S. Holloway, "The influence of potential energy surface topologies on the dissociation of H_2," *J. Chem. Phys.* **93**, 2859 (1990).
7. M.D. Feit, J.J.A. Fleck and A. Steiger, "Solution of the Schrödinger equation by a spectral method," *J. Comp. Phys.* **47**, 412 (1982).
8. J. A. Fleck, J.R. Morris and M.D. Feit, "Time-dependent propagation of high energy laser beams through the atmosphere," *Appl. Phys.* **10**, 129 (1976).
9. E. J. Heller, "Time-dependent approach to semiclassical dynamics," *J. Chem. Phys.* **62**, 1544 (1975).
10. D. Winkler and S. Kamins, **Hypercard 2.0—the Book** (New York, Bantam Computer Books, 1991).
11. D. Winkler and S. Knaster, **Cooking with HyperTalk 2.0** (New York, Bantam Computer Books, 1991).
12. Waite Group, **The Tricks of the Hypercard Masters** (Hayden, 1991)